W9-CBS-882

RAISING SHEEP
THE MODERN WAY

PAULA SIMMONS

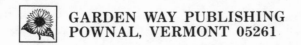
GARDEN WAY PUBLISHING
POWNAL, VERMONT 05261

Designed by David Robinson
Cover by Ken Braren

Copyright © 1976 by Storey Communications, Inc.

The name Garden Way Publishing is licensed to Storey Communications, Inc.
by Garden Way, Inc.

All rights reserved. No part of this book may be reproduced without permission in writing from the publisher, except by a reviewer who may quote brief passages or reproduce illustrations in a review with appropriate credit; nor may any part of this book be reproduced, stored in a retrieval system, or transmitted in any form or by any means — electronic, photocopying, recording, or other — without permission in writing from the publisher.

Printed in the United States by Capital City Press.

Thirteenth Printing April 1984

Library of Congress Cataloging in Publication Data

Simmons, Paula.
  Raising sheep in the modern way.

  Bibliography: p.
  Includes index.
  1. Sheep.   I. Title.
SF375.S56   636.3'08   76-44530
ISBN 0-88266-093-4

CONTENTS

DEDICATION

To Mr. and Mrs. Albert Lund, whose Shepherd *magazine has been a monthly instruction book for over fifteen years.*

To my husband—shepherd, shearer, spinner and weaver—who has encouraged me and made helpful suggestions, and corrected my excesses and errors.

And to Lillian Godbe, who patiently typed the manuscript, with great interest in its content.

AUTHOR'S NOTE

Modern sheepraising has shown a real trend toward the small situation, with emphasis on profitable self-sufficiency. With more and more people keeping a few sheep, the average number of sheep on small farms in this country is similar to the size of the average farm flock in Switzerland—seven sheep.

This small number per flock makes it more urgent that there be no losses due to disease or neglect, and this requires greater knowledge of the fundamentals of sheep health, and of preventive care, and of the latest in medical treatment, should that become necessary.

CHAPTER ONE

STARTING WITH SHEEP

It is preferable to "grow" rather than "buy" into sheep. Starting small gives you the opportunity to get low-cost experience. If you start with fewer sheep than your pasture will support, you will be able to keep your best ewe lamb each year, for a few years at least. And after a few years, as any of your purchased ewes reach an unproductive age they can be replaced by keeping your best lambs.

If you haven't any preference of breed, consider the predominant one in your area. It is likely to be well suited to the climate, and buying close to home saves shipping costs and a stressful ride for the animal. You can also get replacement rams more easily, even trading with other breeders nearby, when you have used yours for a while and want to avoid inbreeding.

PUREBREDS

A purebred possesses the distinct characteristics of a specific breed, and either is registered or eligible for registry in the association of that breed.

The advantages in purebreds are greater uniformity in appearance and production, and a chance of income from the sale of breeding stock. In theory, they may be in better health, as the owner probably would take better care of a valuable animal.

The disadvantages are higher initial cost, plus the cost of registering each lamb, with no better price from the sale of either the meat or the wool than if they were not registered. Also, the financial loss is greater if one dies.

Often it is a good investment to buy a purebred ram to use on grade ewes, for his good characteristics should show up in every lamb that is born. (More on this in Chapter Six.)

Sheep that have the characteristics of a particular breed, without the registration papers or the assurance of their breeding, are called *grades*.

WESTERN AND NATIVE EWES

In predominately sheep-raising parts of the country, some sheep are classified as "native" sheep, and some as "western" or "range" sheep. The ones called native are mostly meat-type animals, large-sized, prolific and usually black-faced. The western sheep are usually fine wool sheep, or a cross of fine wool and long wool breeds. The fine wool sheep were often preferred on the western range, not for their wool, but for their superior herding instinct.

Purebred rams are almost always used on western and native ewes, the breed of ram being determined by the specific market, such as a wool market or a lamb market.

BUYING

It is not always possible to heed the following criteria, because your selection may be limited. But if you know what is undesirable, you can better evaluate the price being asked for the sheep.

Try to avoid the following:

1. Any condition resembling pink eye, or any eye damage.
2. Teeth missing. A sheep with missing teeth can't eat well and may require special care and feeding.
3. Lower jaw not matching upper jaw properly, either overshot or undershot.
4. Lumps in the udder. These may indicate mastitis, so that a lamb would require supplemental bottles, or be a complete bottle baby.
5. Limping sheep. This may indicate hoof disease.
6. Untrimmed feet, turned up at the toes like skis, or overgrown and turned under at the sides.
7. "Bottle jaw." Lumps or swelling under the chin, usually caused by severe internal parasite infestation.
8. Extremely thin ewe. Unless she has just raised twins or triplets, she may be diseased or have heavy load of parasites.
9. Extremely fat ewe. She may not breed, or if bred she may have trouble lambing.
10. Wool going too far down on legs. It is more trouble to shear.
11. Wool covering face. Shearing the face takes time. Wool blindness inhibits eating and mothering. In addition, Australian tests have proven muffle-faced ewes to be less fertile and productive.
12. Fine-boned sheep. Those with larger bones are more hardy and productive.
13. Small size. Potentially smaller production of both lambs and wool. Those that are undersized for their particular breed are not recommended.
14. Narrow or shallow-bodied. Their lambs lack good carcass conformation.

Open-faced sheep (left) has been found to be more fertile and productive than muffle-faced ewe (right).

15. Old sheep. Those over seven or eight years old are progressively less profitable. Even if they have good teeth, they are a poor investment *unless* their price is low.
16. Ragged, unattractive fleece. This may indicate sheep is scratching itself because of ticks or mites. Part the fleece in several places and look closely.
17. Sheep with runny droppings. May be caused by lush seasonal pasture, but also could be caused by internal parasites. Ask how recently they were wormed, and what drug was used.
18. Sheep that were single births. Twinning is quite hereditary and more profitable.

SHEEP AGE *VS*. PRICE

The age of the sheep is important in relation to the *asking price*. Just how many more fleeces and lambs can this ewe be expected to produce? If she is quite old, how much additional feed will she need to compensate for her poor teeth?

TEETH

You can tell a sheep's age by its teeth, but only up to a certain age. As shown in these photographs, a lamb has eight small incisor teeth until it reaches approximately one year of age. Each year thereafter, one pair of lamb teeth is replaced by two permanent teeth that are noticeably larger. By the time a sheep is four years old, all the lamb teeth have been replaced with permanent teeth, and it is no longer possible to tell its age *accurately* by the teeth. You can only estimate by the condition of the teeth. (In case you wondered, sheep also have twenty-four molars.)

Teeth wear drastically shortens a sheep's life. The incisors are all on the bottom jaw, and as the teeth wear down, the amount of tooth below the gumline (about ½ inch) is gradually pushed out to help compensate for the wear. This is partly why the teeth of an old ewe look so much narrower.

The wider part at the top of the tooth is being worn back toward the narrower center-part of the tooth. The tooth is also being pushed up a bit, and the pushed-up part is from below the gumline, and narrower. The gaps between the teeth reduce the efficiency of the ewe's bite. If you listen to an old ewe grazing, you will hear a squeaking of grass as it slips between her narrowed teeth.

Teeth are an indication of the age of sheep. (Michigan State University Cooperative Extension Service)

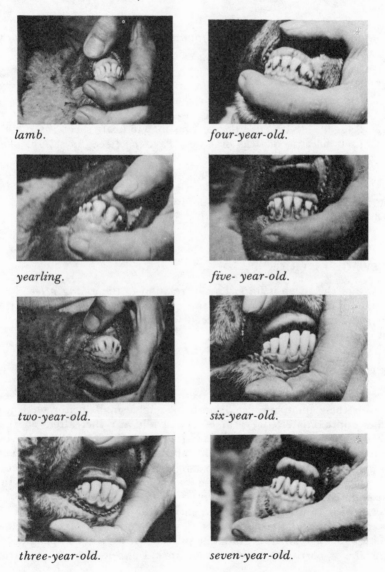

lamb.

four-year-old.

yearling.

five- year-old.

two-year-old.

six-year-old.

three-year-old.

seven-year-old.

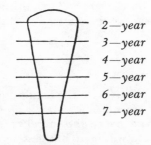

Approximate annual wear of sheep's teeth.

On very low or overstocked pasture, teeth wear down faster from soil and sand. The closer to the soil sheep graze, the more dirt and sand they ingest with their food, and these wear down their teeth like sandpaper. On short pasture, ewes also must take more bites to get every pound of grass they eat, and this wears on the teeth.

SHEEP ACCORDING TO THEIR TEETH

Solid Mouth	Having all adult teeth in place (up to about four years of age).
Spreaders	These are older. Teeth show wear with the under-gum portions, which are narrower, moving up into position.
Broken Mouth	These have some teeth missing. You may get one or two seasons of lambs from them.
Gummers	Sheep that have lost all of their front teeth. A very poor buy.

I would add that gummers may do better than a badly broken mouth sheep, as the gums harden so that they can still chomp off grass, while broken mouths with two or three teeth cannot get a good grip on the grass. If you have an old ewe with a broken mouth who is down to one or two loose teeth, and you are determined to keep her, those few teeth should be pulled with pliers, as they prevent the lower gum from making contact with the upper pad. When the gum hardens, she can eat grass more efficiently than with only a few teeth. She will still need special attention in feeding.

British magazines have reported an orthodontist who has designed dentures for sheep. And in France, Dr. Gilles Raoult, a dental mechanic, is making false teeth for both cows and sheep, charging about $40–$45 per animal. These extend the life of a ewe by at least five productive years, which would be worth it for a twinning ewe with a good fleece.

If you have a dental-student friend who wants to make dentures for your old ewe, have him look first at the teeth of a young ewe. Notice that these front teeth are gouge-shaped, concave without and convex within. This is one reason why sheep can crop the grass closer to the ground than other farm animals. (Another reason is the narrow, flexible muzzle, which is divided by a vertical cleft.)

BUYING OLDSTERS

You can get started in sheep raising with the least outlay by purchasing old ewes, someone else's culls. Frequently they may have been their best ewes. Their years are numbered, and hence the initial cost will be low. If you keep the best of their lambs, you will be in business.

A commercial grower will consider a ewe old at seven or eight years of age, while with good feed, her lambing capability goes on to ten or twelve years. She will do better for you than for her former owner, if she does not have to compete with younger ewes for food. So, you will be doing the sheep a favor, and her owner will probably be happy to find a sale for her.

PREGNANCY OF OLDSTERS

Don't *over* feed them in the early months of pregnancy, as they will need to have better feed in the latter part of pregnancy to avoid toxemia. Encourage

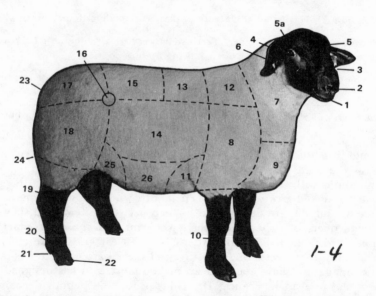

(1) mouth. (2) nostril. (3) face. (4) eye. (5) forehead. (5a) poll. (6) ear. (7) neck. (8) shoulder. (9) brisket. (10) foreleg. (11) foreflank. (12) top of shoulders. (13) back. (14) paunch. (15) loin. (16) point of hip. (17) rump. (18) thigh. (19) rear leg. (20) pastern. (21) dew claw. (22) foot. (23) dock. (24) twist. (25) rear flank. (26) belly. Numbers 18 through 24 represent the leg of mutton. (Suffolk Sheep Association)

them to exercise, since that contributes to their good health. Their lambs may need at least a supplemental bottle feeding (with lamb milk replacer, see details in Chapter Ten) once or twice a day, if the ewes' milk supply is not ample. Start the bottle feeding the day the lambs are born, to insure that they will continue to accept the bottle. Since you are only their extra mama, it is not as confining a chore as if they were real bottle lambs, with no sheep mama.

AVOID SHIPPING FEVER

To prevent shipping fever, we give a shot of Combiotic before transporting sheep. Move them in mild weather, if possible, and avoid rough handling and overcrowding in transporting.

AVOID PROBLEMS

Feed the same type of grain at your farm as the sheep you bought were accustomed to eating. Ask the owner what kind; if it is not easily available, buy some from him. Then gradually change the sheep from their accustomed grain to whatever you intend feeding.

To avoid scours or bloat, sheep should be given their fill of dry hay before being turned out on a pasture more lush than they had before.

Ask the owner when the sheeps' feet were trimmed last, and see if he will trim one of them while you watch. This is an easy way to learn how it is done, and how the feet should look when properly trimmed.

You should know how recently the sheep were wormed, and what drug was used. If they are due for a worming, maybe the owner will do it while you watch.

CHAPTER TWO

SHEEP BREEDS

Sheep breeds were developed in accord with certain needs both in respect to market and the conditions under which they had to be raised. However good the particular breed sounds, it may not be the best choice for your situation. Such things as climate and pasture, how much time and money can be invested in their care — all have to be considered.

Some breeds have a higher incidence of multiple lambing, which is fine if you are able to give them sufficient attention to insure survival and good growth. Twins and triplets, without supplemental grain feeding, will grow more slowly than singles.

Some breeds can be managed to lamb more than once a year but out-of-season lambing does not usually correspond to the best time for pasture grazing, so both ewes and lambs will need special feeding.

More recent breeding programs have placed less emphasis on visual appearance, and more on carefully measured productive characteristics, such as food conversion and weight gains, fast growth of lambs, prolificacy, and clean weight of wool.

A chart at the end of this chapter (p. 42) was prepared by the *National Livestock Producer* in the fall of 1975. Seven sheep experts across the country rated the major sheep breeds for twelve different traits, and then gave them a final breed ranking by tallying them up on a point system. A few traits were missing, and another chart has been prepared by Robert M. Jordan of the University of Minnesota, which lists some of these other traits, with his evaluation.

The main value of these charts is not to see which breed is in first place or second place, but to show a fairly unbiased listing of the strong points and weaknesses of each. This makes possible a more informed choice of which is suitable for your own needs.

In a situation where it is not possible for someone to be at home during the day in lambing season, then "ease of lambing" would be more important than "growth rate." And if you live in a climate of extremely hot summers, then "heat resistance" would be more urgent than "longevity."

If you intend to spin your own wool, you'll have different needs than the person who is primarily interested in locker lambs.

BARBADOS BLACKBELLY

The Barbados is a dark tropical hair sheep, originally from the Island of Barbados in the West Indies and said to have developed from West African stock. Some of the recent interest centered around them comes from their trait of lambing almost twice a year. They are prolific, hardy, and breed out of season. The ewes are nervous around strangers, protective of their young, and need high fences if worked much. Young ewes breed prior to one year old, usually at five to seven months of age, and are very good mothers.

The breed is reputed to be resistant to internal parasites, and field trials have confirmed this, showing 236 parasite eggs/gm/feces compared to 2,300 or 2,077 or 1,490 of other breeds and crosses (Lemuel Goode, N. Carolina).

The North Carolina Experiment Station has been crossbreeding since 1971, using Barbados x Dorset as well as Dorset x Landrace, to improve overall productivity of ewe flocks by raised reproductive performances, plus resistance to heat stress. Some areas of North Carolina have serious problems caused by heat, including ram infertility, failure of ewes to exhibit estrus, fertilization failure, early embryo death, and impaired fetal development.

Blackbelly ram and ewes. (Dr. Lemuel Goode, North Carolina Experiment Station)

Projects are also going on in crossbreeding for year-around lambing ability, the best documented one being by Glenn Spurlock of the University of California. There the average lambing interval has been from 6½ to 7 months, although in multiple lambings (three or four lambs) this interval is about eight months, multiple lambing appearing to retard the return to breeding.

Between 250,000 and 500,000 of the Barbados are now in Texas, many being crossed with Rambouillet, and some crossed with the European Moufflon which is a wild sheep, and used for hunting on game ranches. Since the Barbados withstands both heat and cold very well, it is especially suited for Texas.

Some of the Barbados in this country have horns, usually a result of some prior crossing. A cross with Dorsets improves wool, decreases nervousness and increases docility. There is some crossing done with handspinning wool in mind, obtaining the dark fleece of the "Barbs" and their accelerated lambing, but crossed to a sheep of a better spinning-quality wool. The crosses then show even more variety of dark coloration than the pure Barb.

CAMBRIDGE

Dr. J. B. Owen of Cambridge University started ten years ago to search for unusually prolific ewes of any breed, ones who had produced three consecutive sets of triplets. Of this collection, many were Clun Forest ewes, but also Llyn and Llanwenog ewes from Britain, and other Welsh-breed types. These were then bred to Finnsheep rams, and the final result was a breed that has lambing percentages of 300 to 400 percent.

The eight farmers who are cooperating in this experiment have formed a Prolific Sheepbreeders Ltd., and have about 1,000 ewes of this new breed, tentatively called the Cambridge. In addition to the "litter" lambing, the carcass of the breed is comparable to that of the average meat breed, a considerable improvement over that of the Finnsheep.

CHEVIOTS

There are two distinct types of Cheviots, with the Border or Southern Cheviot being the smaller size and was improved by selection from the original stock rather than by crossbreeding, and is the predominant type of Cheviot in this country.

The Cheviot started as a mountain breed, native to the Cheviot hills between Scotland and England. It is extremely hardy and can withstand hard winters, and graze well over hilly pasture and in high altitudes. It lacks the herding instinct needed for raising on open range, but does well in a farm flock. On scant pasture, the Cheviots spread out and get all the available feed.

They are active and high strung, being alert both in appearance (erect pointed ears) and behavior. They are good mothers despite their nervousness, and their newborn lambs are more hardy than many other breeds. Because of their small head, they experience few lambing difficulties, and they raise a good meat lamb. They are short and blocky, with clean faces, strong noses,

Cheviot ewes. Breeder is Virginia Rowell, California.

and black nostrils and lips that combine with the sharp ears to make them a very recognizable breed. They have a lightweight fleece, medium wool about 48s to 56s that is easy to use in handspinning.

The "other" Cheviot is the North Country Cheviot, a larger size animal, more of a Scottish breed, and has a more pronounced Roman nose. Its size is the result of earlier crossbreeding.

CLUN FOREST

This is a new breed in this country, and not well known yet. The first six ewes were brought from Northern Ireland to New York in 1959, but the first large importation was not until 1970, from Shropshire, England. Because of strict quarantine regulations, no breeding stock from this shipment could be sold until 1973.

Clun Forest ewes are prolific, almost always having twins. With narrow sleek heads and wide pelvic structure, they show unusual ease of lambing without assistance. With the forty ewes imported in 1970, assistance was only given to one ewe in three years of lambing. The owner, Mr. Turner, afterward said that it might not have been necessary (she was having triplets). They have a strong mothering instinct, even when lambing as yearlings. An average of 80 percent of Clun ewe lambs will breed as lambs at eight or nine months old, and lamb as yearlings.

Geoffrey Bowen, in *Wool Away*, claims that Clun Forest twin lambs easily attain 45 pounds dressed weight at four months of age, showing the fast growth of these lambs, attributed to the richness of the ewe's milk—not more fat content than other sheep, but more protein.

Clun Forest ewes, each with twins under one week old. Breeder is Mrs. Warren G. Menhennett, Cochranville, Pa.

They are adaptable to all climates, being successfully grown from the rainy areas of England to the hot dry climate of Ghana, in low, marshy places of England to high hills in Wales and Scotland. They are good foragers.

Another quality that makes them valuable is their longevity, bearing twin lambs and good fleeces to age ten and twelve. Their medium wool is of 58s count.

The Clun Forest Breeders Association members have voted unanimously to sell only rams for breeding purposes that have been born a twin, triplet or better.

COBB 101

This is the one that is called the "broiler lamb," and is a new British breed that was inspired by poultry raising. It was developed by the Cobb Breeding Company, which had previously only dealt in poultry—the Cobb meat-type chicken. It is a combination of Finn rams and British meat breed ewes, with complicated computerized records of performance. The ewes are docile, the rams highly virile (semen-tested), with a sleek head and small horns. The ewes are mated at seven to eight months of age, and lambs reared on cold-milk formula, in pens, then on pelleted growing ration. They attain 80-pound weight in twelve weeks. Ideally, they will have two lamb crops a year.

COLBRED

The Colbred is a British records-derived hybrid developed by Oscar Colburn, again by application of poultry genetics to sheep breeding. It has a combination of desirable traits achieved by crossing East Friesian, Border Leicester, Dorset Horn, Clun Forest and Finnsheep breeds, all prolific and good milkers.

COLUMBIA

The Columbia is an American breed, developed since 1912. Started in Wyoming, it was sent to the U.S. Sheep Experimental Station in DuBois, Idaho, in 1917. It is the result of a Lincoln ram and Rambouillet ewe cross, with interbreeding of the resulting crossbred lambs and their descendants without backcrossing to either parent stock. The object of the cross was to produce more pounds of wool and more pounds of lamb. Their large robust frame and herding instinct have made them excellent for Western range purposes, but they have also proved admirably adaptable to the lush grasses of small farms in all parts of the country. Heavy wool clip, hardy and fast growing lambs, open faces, and ease of handling are characteristics appreciated everywhere.

They have medium wool, in the 50s and 60s range, but predominately about 56s. It has light shrinkage, and is an excellent fleece for handspinning.

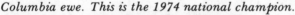

Columbia ewe. This is the 1974 national champion.

It is an all-white breed, polled, and open faced. Since it is such a relatively new breed, even more improvements are expected, and there are Columbias being raised at a number of experiment stations, including the Washington State University Research Center at Prosser.

COOPWORTH

The Coopworth is a new breed in New Zealand, a cross of Border Leicester x Romney, and has unusually strict registration requirements. Performance recording is mandatory. For "merit" registration, a ewe must have raised six lambs by the age of 4½ years, and be above average for yearling fleece weight and total weaning weight of her lambs. If she fails to rear two lambs on more than one occasion after qualifying for the merit register, she loses the merit status.

A ram must be of multiple birth, and born within the first twenty-one days of lambing. He must be out of a merit ewe or one with a lifetime 170 percent lambing record. His sire has to be at least 2.2 pounds above the average of his contemporaries for adjusted weaning weight* and 0.22 pounds above average for yearling fleece weight.

CORRIEDALE

The Corriedale is a Merino-Lincoln cross, developed in Australia and New Zealand and first brought to this country (Wyoming) in 1914.

Its dense fleece is medium fine, 56s grade, with good length and softness and light shrinkage, somewhat between medium wool and long wool, a favorite of handspinners in many areas of the country.

Its face is clean of wool below the eyes, and the sheep is hornless. Bred as a dual-purpose sheep, it has good wool and good meat for greater profits, and is noted for a long productive life, which means greater return on your investment. Because of a marked herding instinct, it is also a good range breed.

COTSWOLD

The Cotswold has very long coarse wool, 8 to 12 inches or more, that is wavy, hanging in pronounced ringlets, with wool hanging over the forehead. There is very little shrinkage in this fleece, but it is sometimes excessively hairy on the thighs.

The Cotswold thrives on dry rolling land, and is similar to the Lincoln and Leicester. It is a very large sheep, whose most characteristic feature is the long tuft of wool hanging over the face.

---

*All lambs reared as singles are excluded when establishing the adjusted weaning and fleece weight.

Corriedale ewe and twins. (Shepherd *magazine*)

Cotswold ewe. (Sheep Breeder and Sheepman *magazine*)

DEBOUILLET

Development of the Debouillet was begun by Amos Dee Jones in New Mexico in the 1920s. This was a cross of Ohio Delaine Merino rams and Rambouillet ewes, the successful crosses showing the length of staple and character of the Delaine fleece and the large body of the Rambouillet. By 1927 the ideal type was attained, and a line breeding program begun. The breed was registered in 1954, starting with 231 rams and 1,587 ewes.

Debouillets are open-faced below the eyes and over the nose, have a good belly wool covering, and shear a heavy fleece of long staple fine wool. Rams can be horned or polled. Even under adverse conditions, the ewes produce desirable type market lambs of excellent weight.

Debouillet lambs that are eligible for registration by bloodline must be one year of age and in full fleece when inspected by an association inspector. Wool must be 64s grade or finer, with staple of 3 inch minimum, and deep close crimp.

DORSET

The Dorset Horn ewes and rams have horns. The Polled Dorset are bred to be without them, on both sexes. The first Polled Dorsets were developed at North Carolina State College and first registered in 1956.

Dorsets originated in England, but their history is not well known, although it is believed that the breed developed more by selection than by crossbreeding.

The Dorset has very little wool on the face and legs and belly, and a light-weight fleece that is good for handspinning.

This sheep has a large coarse frame, with white hooves and pink skin. The ewes are prolific, often twinning. They are good milkers, having even been kept in dairies at one time in England, so their lambs grow well. They are good mothers, and have a special feature of early breeding for fall lambing, and it is even possible to lamb twice a year. A *Shepherds Guide* of 1749 described Dorset Horn sheep as "being especially more careful of their young than any other."

A Dorset ewe has a good appetite and good digestion, can take a lot of feed and feed her lambs well, to get them to market by the time many other breeds are just lambing. These out-of-season lambs can command good prices.

DRYSDALE

This is the New Zealand Romney with the "carpet wool gene," a genetic mutation that has happened in some Romney sheep there. It was first discovered by Dr. F. W. Dry of Massey University in the 1930s. A lamb with that N gene can be identified by a patch of pure hair that is noticeable under the shoulder at birth. This was a very uncommon occurrence, and by

Dorset ewe with lamb. (Continental Dorset Club)

considerable genetic research, breeding and selection, was finally developed into an established breed.

Flocks have remained closed since 1967, with a ban on exportation. Overseas Carpets Ltd. distributes and controls the rams, to protect the valuable output of their use, and farmers raise them under contract.

The wool produced is a hairy 28–36s count, with a 15–inch growth in eleven months. It grows so fast that some are sheared every eight months. This uniform grade of carpet wool is valued for the manufacture of resilient carpets with long life, and without undue "tracking" of the surface.

EAST FRIESIAN

This is a German breed, raised primarily for milk, particularly for cheese production. It is the highest milk producer of all the European breeds, as well as being very prolific. While it is a large-sized sheep and lambs have a good growth rate, it does not have a good meat carcass.

It shears a heavy fleece of 48–50s wool.

In crossbreeding for increased milk production to accompany prolificacy, this would be a valuable genetic addition. In southeastern France, this is one of the three milk breeds crossed together for the famous supermilking sheep that are the basis of the Roquefort cheese production there.

FINNSHEEP

The Finnish Landrace is native to Finland, fairly new in this country, but a fast-growing breed. It is valuable for crossbreeding to introduce its unusual prolificacy, being called the sheep that "lambs in litters." Up to ⅛ Finnsheep in a cross will increase lambing percentages somewhat, although the same thing can be done with good culling and breeding for twins and triplets in your own stock.

Finnsheep ewe and quintuplets. This remarkable ewe, Langelmaki 57A, had a total of twenty-seven lambs in seven years, including these quintuplets which she raised without assistance. Her owner is Werner Grusshaber of Finnlamb Farm, Brantford, Ontario.

When these sheep have quadruplets and quintuplets and sextuplets, it is customary to leave two or three on the ewe, and feed the rest on lamb milk replacer, after they get the colostrum. They are good mothers, and topped all the breeds in the "breed chart" for ease of lambing.

First brought to the United States about 1966, they have expanded rapidly, and more than 2,600 were registered as Purebred Finnsheep by 1975, with many more not registered. This breed will register black sheep, and has a fairly high incidence of black lambs, for this feature has not been purposely eliminated in the breeding.

The ewes will breed young, most lambing before they are a year old. Ewes that are to be bred at the age of six-seven months should weigh at least 100 pounds before breeding, and at least 125 pounds at lambing. They will need special feed and care during gestation, to meet the nutritional needs of their multiple lambs. If these young ewes are to grow and reproduce at the same time, the breeder must give extra care.

Finnsheep have naturally rather short tails, which don't always require docking.

FINNSHEEP CROSSES

There is widespread use now of Finnsheep for crossing with other breeds to increase their lambing percentages. In commercial flocks, the meat carcass

This Finnsheep Cross ewe is half Dorset, half Finn Cross. She raised these quadruplets without help, and she alternates between triplets and quadruplets, showing the success of the half Finn Cross. (Mary Sue Ubben, Breinigsville, PA)

of the quarter Finn crossbred lambs appears to be indistinguishable from that of the meat breed used in the cross, while greatly improving lambing percentages.

HAMPSHIRE

The Hampshire is one of the largest of the medium wool meat-type sheep with very rapid growing lambs. While they do not do well on rough or scanty pasture because of their size and weight, they do nicely on good pasture and the lambs can usually be marketed off grass. The ewes are good milkers and fairly prolific, but do not always lamb easily, probably due to the large head and shoulders of the lambs, which are quite heavy at birth.

The Hampshire has a large head and ears, is hornless, and the dark face is closer to a rich dark brown than to black, with legs of the same color. Its fleece is light weight, medium wool (48s to 56s) and fairly short.

The rams are one of the favorite breeds to use on fine wool range ewes, to attain the fast growth of market lambs. The rams are also noted for their keen sense of smell, important in detecting estrus in the ewes.

Lively Hampshires head for the pasture. (John D. Wibbels, Jeffersonville, IN)

IMPROVER

Brian Cadzow of Glendevon, Scotland, has the most internationally famous and successful new breed, as well as the least known in this country. His crossbred rams are tailored to order for special conditions in particular countries. He started in 1962, on his 1,000-acre farm, drawing from a genetic pool of nine pure breeds, including Ile de France (good carcass) and Finnsheep (prolific).

The Improver has variations within the breed, depending on the requirements of the customer and the type of native ewes that need improving.

The drug firm of Chas. Pfizer has taken over foreign distribution, and rams have been shipped to Canada, Hungary, Spain, Belgium, Kuwait, Germany, Italy, Saudi Arabia, Bulgaria, Iran, Czechoslovakia and Britain. They are crossed with native breeds of these countries to give more milk, heavier lambs, better growth, and lambs which can be weaned at eight weeks. They are especially valuable to increase sheep productivity in the Middle East countries, where sheep are an important food source.

Part of the success of these sales is due to the ongoing advisory service that goes with the sheep, giving constant advice on management, breeding, nutrition, veterinary care, parasite control, and housing.

The Glendevon station has over 6,000 sheep of the nine grandparent breeds, with computer data analysis for the production of new hybrids for specific needs.

JACOB SHEEP

The Jacob sheep, once thought to be in danger of extinction, has now become so popular and relatively common in England that they have their own breed society, the Jacob Sheep Society, Jenkins Lane, St. Leonards, Tring, Hertfordshire.

Their four horns and mottled fleece give them a unique and distinctive appearance, but their appeal in recent years has been for handspinning wool.

KERRY HILL

This British breed originated in the Kerry Hills of Montgomeryshire, early in the 1800s. It is a hardy sheep and does best in hilly country, but is raised successfully even in lowlands and marshy areas.

The Kerry Hill has black and white markings on both face and legs, and a very black nose. The ewes are prolific and robust, and the rams noted for their olefactory powers, important to detect estrus in ewes who show only faint signs, as in out-of-season breeding. The rams are used in Wales for crossbreeding with the Welsh mountain ewes, producing a speckled face sheep that is very popular there. It is a good all-purpose breed for both meat and wool. See photo, p. 39.

These Jacob sheep and ram were the first imported into Canada, by R. A. K. (Tony) Turner. He took them to Nova Scotia in October 1975.

LEICESTER

The English Leicester has wool over the crown of the head, and resembles the Lincoln, except it is smaller, with a wool tuft on its forehead.

The Border Leicester has no wool on the head, and less depth of body, a definitely Roman nose, and stylish look. Their fairly erect ears seem far back from the wedge-shaped face. These are long-wool breeds, with coarse curled wool, 40s to 48s grade.

LINCOLN

The Lincolns are from Lincolnshire in England, and are the largest of the sheep breeds, and slow maturing. Their long fleece is dense and strong and heavy, and they have forehead tufts. The breed is fairly hardy and prolific, but the lambs need protective penning for the first few days.

The Lincoln is not an active forager, and is best adapted to an abundance of pasture and supplements.

They do not stand cold rainy weather too well, as their fleece parts down the middle of the back, allowing the cold to hit their backbone, a sensitive area on sheep. Their fleece, however, is resistant to the deterioration shown in the wool of other breeds when parting along the back. It is sought by handspinners for the special long-wearing qualities and lustrous appearance. We like the wool for handspinning into almost indestructible sock yarn, and blended with other wools to make a strong weaving warp, with an attractive sheen.

This is J. D. Paterson's champion Border Leicester ram at Ellesmere. (New Zealand Farmer *magazine)*

The author's favorite sheep, Mary, a charcoal-gray Lincoln.

While there are relatively few purebred Lincoln herds in this country, the breed has been extremely important in the crossbreeding to develop a number of breeds.

MEATLINC

This is another British achievement, with record keeping by computer. The Meatlinc was developed by Henry Fell at Worlaby in Humberside, England. By a four-way cross of two British breeds (Suffolk and Dorset Down) and two French breeds (Ile de France and Berrichon du Cher) he has attained a fast growth rate, low fat cover at higher lamb weight, and a high flesh-to-bone ratio. This new breed is expected to be most valuable as a sire breed, to transmit these growth and carcass traits to the lambs.

MERINO

The Merino sheep, so famous for their fine wool, originated in Spain. They are descended from a strain of sheep developed during the reign of Claudius, from 14 to 37 A.D. These Tarentine sheep of Rome were crossed with the Laodicean sheep of Asia Minor by the Spanish.

Their fleeces are heavy in oil, and lose much of their weight in washing. Since their lambs are small and slow-growing, the main income is from the fleece.

There are three types of Merinos now. Type A has folds in the skin. The wool that grows in the folds is much coarser than the wool on the rest of the body, making more variation in individual fiber sizes on this than in any other breed. The folds and wrinkles of their skin were originally encouraged to give more skin area, and to thus have more area of wool follicles, and more wool production.

This wrinkling feature of the Merino is no longer considered desirable, and the Type B Merino has fewer folds. The Type C Merino has even fewer folds and wrinkles, and is known as the *Delaine Merino,* or *American Merino.*

The lambs of the Delaine Merino grow a little faster than those of Type A and B, but the ewes are not especially good milkers.

The Merino has the typical fine-wool trait of herding closely, and they can travel far for feed and water, so are good on open range. Many of the Delaine Merino are raised in Texas.

The German Saxony Merino has even finer wool than these Spanish Merinos, and was at one time raised in great numbers in the Northeastern states, and as far west as Ohio.

MONTADALE

The Montadale is an American breed, originating about 1932 in the St. Louis area, a cross of Cheviot rams and Columbia ewes.

The small head eliminates many lambing problems, and they are prolific lambers and good mothers.

Merino sheep, raised in Vermont. (Vermont Historical Society)

Champion Montadale ram.

They have heavy fleeces with little shrinkage, open faces, and clean legs. Wood grade is ⅜ blood.

The Montadale Breeders Association was organized in 1945, and by 1974 there were more than 68,000 registered.

The Montadale is easy to recognize, with a lot of style, a beautiful face, and those alert Cheviot-like ears.

THE NO-TAIL SHEEP

South Dakota State University spent fifty years (1915–1964) on a breeding experiment to develop a no-tail breed, before its disbandment. They originally started with two imported Siberian Fat-rumped tailless rams — bred with Hampshire, Cheviot and Shropshire ewes in the first cross. Later breeding also involved Southdown, Columbia and Rambouillet sheep. It was not until the fifth generation that a lamb was born without a tail.

The original rams had spiral horns, were wild and nervous, had fleece with fine under wool and long guard hairs, but were hardy and good grazers. Ewes of that breed are excellent milkers and mother their lambs well.

The goal of the breeding — to develop a high producing and tailless sheep — was not completely accomplished. In the later years about 40 percent of the lambs were really tailless at birth, another 20 percent had tails of 1 inch or less, and the remaining 40 percent had tails ranging up to 6 inches in length.

The tailless genetic trait did not become completely fixed in the flock, and at the time of the termination of the project, it was stated that the

No-tail ram at South Dakota State University, Brookings, S.D.

Oxfords, owned by Mrs. Dan Korngiebel, Cuttingsville, Vt.

"flock offered no other performance superiority not already available in our more widely distributed sheep breeds." It was not sufficient to be "as good" as existing breeds. The object of a new breed is to be "better."

However, in fifty years they were bred into sheep that are hornless, with ¼ blood fleece having no evidence of hairiness, and medium-sized breed with good mutton type. These resulting ewes were good mothers, good milkers, hardy, but not exceptionally prolific, and still very nervous.

This certainly demonstrates the power of selection and crossbreeding to produce genetic changes over a period of time. Not a new breed now, but a most noteworthy project.

OXFORD

The Oxfords originated in Oxford County, England, where they were bred from a primarily Cotswold and Hampshire foundation, which makes them a large heavy breed, with a good fleece weight and good length of wool (med. wool, 46s to 50s). The fleece is valuable for handspinning. They were successful in combining the hardiness, muscling and wool quality of the Hampshire with the great size, growthiness and wool quantity of the Cotswold, and were first recognized as a true breed about 1862. They were imported into this country as early as 1846, and in expanding numbers into the 1900s.

The 1930s saw an overall trend toward a more compact type of sheep throughout the industry, and Oxford-sired lambs were criticized for being too large and growthy, and not carrying enough fat at desired market weight. Oxford raisers were pressured to reduce the size of the breed to conform to the popular demand of the times. By the late 1950s, the outlook on fat in relation to lean edible meat had changed, and breeders looked again to larger and more efficient and productive sheep. As not all Oxford breeders had gone along with the reduction in body size, there was available stock to again "improve" the breed by regaining its large size and muscling. These are called *Modern Oxfords.*

It is most valuable as a sire breed. Rams now weigh up to 300 pounds, and in crossbreeding they contribute size and muscling to the resulting lambs. Being easily handled in small pastures, Oxfords are well suited to farm raising, and thrive with good feeding. The ewes are docile and heavy milkers. Since the breed has a head rather small for the body, the lambs are born easily.

Their faces and legs are usually light brown, but anything from light gray to dark brown is now acceptable, and a white spot on the end of the nose is quite common. With only a partly wooled face, there is no tendency to wool blindness.

PANAMA

The original Panama stock is the reverse of the parent breeds of the Columbia. Breeder James Laidlaw wanted to develop a sheep to replace the small Merinos that were very common in Idaho, and mated Rambouillet rams to Lincoln ewes. The aim was to get a more rugged sheep, with finer wool and better herding instinct than from the opposite cross of Lincoln rams on Rambouillet ewes. He felt that the ram had more influence on the offspring than the ewe, though this is still a matter of some controversy.

The first cross was made in 1912, starting with 50 rams and 1,600 Lincoln ewes. With this large number of rams and ewes, the breeders were able to

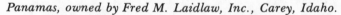

Panamas, owned by Fred M. Laidlaw, Inc., Carey, Idaho.

avoid the inbreeding problems that sometimes arise in the attempt to form
a new breed. The Rambouillet rams were only used for three years, and
after that only Panama rams were used. By the end of five years, 1,000
Panamas were selected for the herd, and the remaining Lincoln ewes were
sold off. After the first few years, the breeders started selecting for polled
rams, and soon the breed was pure polled (hornless) as it is today.

The American Panama Registry was started in 1951, and all registered
Panamas must be direct descendants of the original Laidlaw flock. They
are good-sized sheep, hardy, good mothers, good milkers, with heavy fleece.
A ewe fleece weighs 11–14 pounds, and is about ⅜ blood wool grade.

POLWARTH

The Polwarth breed has a history dating back to the 1880s, when it was
bred to meet environmental conditions of Western Victoria, Australia,
where the climate is too cold and wet for pure Merinos. It is a first cross
Lincoln/Merino ewe mated with a Merino ram. Progeny from this are then
mated. It was first known as the *Dennis comeback* and later *Polwarth* after
the county in which it originated, in accordance with English practice.

The breeders association was formed in 1919, and defines Polwarth as a
dual-purpose sheep, approximately ¾ Merino and ¼ Lincoln, inbred to a
fixed type with the emphasis on fleece. Fleece is about 58 to 60 count, not
less than 4-inch length, dense and even, and carried well down on the belly.
The wool-free face eliminates the need for facing (face-shearing is called
wigging in Australia) and prevents wool blindness. Ewes will take rams at
any time of the year, and have been successfully lambed twice a year, with
high twinning. They are excellent mothers, lamb easily, and are good
milkers.

The Polwarth has developed into an important breed in Australia, and
has been exported to New Zealand, China, Nepal, Taiwan, Brazil, Uruguay,
Argentina, Falkland Islands, Peru, Pakistan, South Africa, Kenya, Korea,
and recently to the United States.

POLYPAY

The Polypay is a new breed developed at the U.S. Sheep Experiment Station in Idaho, announced as a breed in the Spring of 1976. It started with
initial crosses of Targhee x Dorset and Rambouillet x Finnsheep breeds.
These crosses were then crossed to form a four-breed cross. The lines were
then closed with intensive within-line selection for lamb production when
given the opportunity to breed twice a year.

It is a superior lamb production breed, with a quality carcass, and is
giving outstanding performance in twice-a-year lambing, and also superior
performance for conventional lambing. Only spring-born ewes which lamb
at one year of age are kept for replacements.

The Rambouillet and Targhee breeding is included to retain hardiness
and breeding season. Dorset breeding contributes to carcass quality, milk-

Polypays. (U.S. Sheep Experiment Station, Dubois, ID)

ing ability and long breeding season. The Finnsheep breeding promotes early puberty, early postpartum fertility and high lambing rate.

Careful and intensive selection has been used to increase further the ability of this new breed to produce at a high level, and the breeder certification program requires a careful mathematical production index be kept on every ewe and ram.

The fleece of the Polypay averages ½ blood wool grade, with an average ewe fleece weighing about 8 pounds, and being very clean wool because of slatted-floor barn confinement during lambings and breedings. Wool weight would be higher if these ewes were not gestating and lactating twice each year.

The Polypay sounds like another success from the same sheep experiment station that perfected the Columbia breed.

RAMBOUILLET

The Rambouillet is the French version of the Merino, developed from 386 Spanish Merinos imported by Louis XVI in 1786, for his estate at Rambouillet, and crossed with his own sheep there. They are very large and strong bodied, with very little wrinkling in the modern Rambouillet, except perhaps a little across the brisket. The fleece is less oily than the Merino, so has less shrinkage.

They are hardy, with a remarkable herding instinct. They graze during

The herding instinct of the Rambouillet is shown here. (American Rambouillet Breeders Association)

the day, and at night they will gather closely to sleep. They are good for open range, and can adapt to a wide degree of climate and feed conditions.

The ewes can be bred early, to lamb in November and December, and the lambs give good yield in boneless trimmed meat cuts.

They are a dual-purpose sheep, with a desirable carcass and good wool production. The rams have horns, and both sexes have white feet and open faces.

This is the breed that came in number one on the breeds chart.

ROMELDALE

The Romeldale is a cross of Romney rams and Rambouillet ewes, producing ½ blood wool with very little shrinkage, making more pounds of clean wool than fleeces of fine wool. The straight Romeldale lamb is very marketable, and also one that can be saved for flock replacement ewes.

This breed is found mainly in California, and its popularity has not spread because of an inactive breed association and registry to publicize it, and also a general unavailability of good Romeldale rams.

Hank Sexton's flock in Willows, California dates back to the 1920s, and his flock improvement system to maximize ranch profits by increasing pounds of quality lamb and wool produced per ewe and per acre has been written up by the University of California Extension service as a model for this type of program. The local Agricultural Extension agent says that Mr.

A Romeldale ewe and her twins. (Hank Saxton flock)

Sexton has increased the ratings of his Romeldales at least ½ point in all classifications, over the breeds chart ratings at the end of this chapter.

The Romeldale is slightly smaller than the Columbia.

ROMNEYS

The Romney is an English breed, there called the Romney Marsh, after a low marshy reclaimed area where they are thought to have originated. They are said to have hooves less prone to the diseases of wet weather, and be somewhat resistant to liver flukes, another danger of damp pastures.

Romneys have a quiet temperament, can do well on a medium-good pasture, but are not suited to hilly country or hot dry climates. They have little herding instinct, but are easily managed.

They have a finer fleece than the other long wool breeds, but long and lustrous, almost as fine as some of the medium-wool sheep. Except for a tuft of wool on the forehead and short wool on the lower cheek, the rest of the head is clean. Their fleece does not have a tendency to part along the back, so they do well in rainy weather. It is an excellent handspinning fleece. Many are raised in Oregon.

They are a good quality meat animal, with a delicate taste. Recent studies have indicated that wool grade and flavor are actually related. Taste tests at the University of Idaho, using lambs raised at the U.S. Sheep Experiment Station in Dubois, found that as breeds go from fine to coarse wool, the amount of "muttony" flavor in the meat decreases as the wool grade gets coarser. This may explain why now the Romney has almost replaced the Merino among the flocks of New Zealand.

Romney ewe and lamb.

RYELAND (RYLANA)

This small English breed has been called the "white-faced Shropshire" and originated in Herefordshire in England. It has good-quality medium wool, and the lambs are fast growing. The ewes have a docile temperament and make good mothers. It is one of the oldest breeds in Britain, and was originally an even smaller size than it is now.

It was one of the breeds used to develop the Australian Poll Dorset.

SCOTTISH BLACKFACE
(BLACK FACE HIGHLAND)

This breed originated as a mountain sheep in Scotland, and it is a hardy quick-maturing meat animal. It has a lightweight fleece of long coarse wool. Both sexes have horns. In addition to an attractive and stylish fleece, its Roman nose and unusual black-and-white face markings set it apart in appearance. The mottled faces are preferred over the solid color black face, in some parts of England, where the markings are said to indicate greater disease resistance.

Ryeland ram. (Sheep Breeder and Sheepman *magazine*)

The distinctive Scottish Blackface. (Sheep Breeder and Sheepman *magazine)*

SHROPSHIRE

Shropshire is one of the "down" sheep, developed in Southern England in the low hills called "downs."

It is a medium-small-sized sheep, with good meat lambs, but needs abundant feed. First known as a fixed breed in 1848, it was imported into the United States in 1855, and became well established as a popular breed. Its wool is $\frac{3}{8}$ to $\frac{1}{2}$ blood, with average fleece weight of about 10 pounds.

The excessive face covering so encouraged in Shropshire breeding in the past has recently become unpopular because of the high incidence of wool blindness, and is now being bred out.

SOUTHDOWN

The Southdown was at one time the favorite meat breed, with a medium small size, but good weight. It has a short broad head, dark legs and face partly covered with wool.

It is one of the oldest English breeds, originating in the South Downs, a hilly portion of Southeastern England, and is well adapted to grazing a hilly pasture.

A small, short and blocky meat lamb is not held in the same esteem that it was at one time, so the number of Southdowns appears to be dwindling, from a high of over 11,000 registered in 1958 and 1959 to only 3,558 in

A modern type Southdown with New Zealand blood lines, owned by George A. Downsbrough, State College, Pa.

1975. Actually, the 1975 figure is up about 200 from 1974, so the numbers may be leveling off now.

The Southdown fleece is a medium wool, 56s to 60s, and its short staple is suitable for handspinning into fine yarn.

SUFFOLK

The Suffolk is the most popular breed in this country, with nearly 47,000 new registrations in 1975. It is a handsome sheep whose black face, ears and legs are free from wool.

The ewes are prolific and good milkers, with very little trouble at lambing. Their lambs grow rapidly, with more edible meat and less fat than many other breeds, plus a fine texture.

Although they placed fourth in the breeds chart, they rated highest in ewe size and ram size, highest in growth rate, feed efficiency and muscling. They were topped only by the Finnsheep in ease of lambing and milking ability, in the breeds on that chart. They led all the other breeds in ewe size, ram size, and muscling. However, they lost points by being far down the scale in longevity and in wool production, having short wool with a fleece that is lightweight.

They are active grazers, able to rustle for feed on dry range, and they travel far to look for feed.

Suffolk ewes. (E. William Hess)

Originally an English breed, it was developed by the breeding of the dark-faced Southdown with the old Norfolk sheep, a black-face horned sheep that was hardy and prolific with meat of a superior texture but with poor conformation. The resulting breed combined the best of these parent breeds, with growing popularity in both purebred herds, and in usefulness for crossbreeding. Suffolk rams are used to cross on Rambouillet range ewes, to obtain the desirable Suffolk qualities in their lambs.

Records from 1920 show the registered Suffolk number at 805 in this country; in 1975 alone over 45,000 Suffolk were registered.

TARGHEE

The Targhee is a hardy American breed, developed by mating outstanding Rambouillet rams to ewes of Corriedale x Lincoln Rambouillet stock, and ewes of only Lincoln-Rambouillet, and following that by interbreeding the resulting lambs. This work was done since 1926 by the United States Sheep Experiment Station in Dubois, Idaho to meet the demand for a breed of sheep that was thick in natural muscling, prolific, producing high-quality apparel-type wool, and adapted to both farm and range conditions. It gets its name from the Targhee National Forest on which the Experiment Station flock grazes in the summer.

It is a large-sized, dual-purpose sheep with a good meat type and heavy

Yearling Targhee ewes from the U.S. Sheep Experiment Station, Dubois, Idaho.

fleece (11 to 16 pounds) of good wool, about ½ blood, 3 inches length or more. It has a clean face and no skin folds, with ewes weighing from 125 to 200 pounds, and rams from 200 to 300 pounds.

Experimental work at the University of Idaho has shown the Targhee to have an inherited resistance to internal parasites and to hoof troubles. The breed can also claim a very long productive life.

It is noted for an ease of lambing, and high percentage of twins or triplets.

On the National Livestock Producer's breeds chart, shown at the end of this chapter, it took second place.

TEXEL

The Texel in Holland is receiving considerable attention from breeders, but because of a long-incubation-period pulmonary disease prevalent in this breed, it does not look promising for importation in the near future. It has many genetic advantages, such as high fertility, large size, rapid growth rate, excellent carcass, and a high wool production of fine crossbreed fleece. It is a hardy all-purpose sheep with good milking ability. Most Texel lamb carcasses from Holland are sold in France.

This breed resulted from many crossbreedings of the native "polder sheep" (grazers on polderland reclaimed from the sea) with British breeds such as Border Leicester and Lincoln to get a better meat animal, which was then crossed back to rams of the original sheep.

Texels adapt easily to new environment, with many selling into South America. While prolific, it lambs only once a year, and is not ideal for open range use as it does not have a good herding instinct.

TUNIS

The Tunis is one of the oldest of the distinct sheep breeds, dating back over 3,000 years. The first importation into this country was in 1799, with sheep from the flock of the Bey of Tunis, in Tunis, Africa, being brought to Pennsylvania. From there they spread mainly to Virginia, Georgia, and the Carolinas.

A Tunis ram was used by George Washington to rebuild his flock, which declined in numbers and vigor while he was serving as president. The Tunis could have been a major breed in this country, had not most of the Southern flocks been destroyed during the Civil War.

They are medium size, hardy, docile, and very good mothers. The ewes are known for breeding out of season, and with proper management they can be bred almost any month of the year. An unusual color of reddish-tan hair covers their legs and faces and their long broad pendulous ears. They have a medium heavy fleece of $\frac{1}{4}$ to $\frac{3}{8}$ blood, very good for handspinning. The lambs are a reddish color when born and gradually lighten to white, although retaining the distinctive red-tan on their legs and face.

The Tunis does well in a warm climate, and the rams remain active in very hot summer weather. Although they are a superior breed for a hot climate, they are raised successfully almost anywhere, and their concentration here is mainly in the Northeastern states.

With the current interest in out-of-season breeding, prolificacy, and milking ability, the Tunis is again increasing in numbers, as well as attracting attention in crossbreeding research projects.

Tunis ewe, with Kerry Hill ewe in background. (Flock of Isaac Hunter, Dowagiac, Michigan)

WARHILL

Crossing experiments conducted in the 1930s by Fred Warren of the Warren Livestock Company and John Hill, then dean of the School of Agriculture at the University of Wyoming, led to the development of the special-purpose twinning Warhill breed.

The Warhill is principally derived from the Panama, Columbia, Rambouillet and Targhee breeds, all being white-faced with strong Merino backgrounds. It has evolved through forty years of intense selection from herds numbering up to 30,000 sheep, under range conditions. Multiple lambing, high milk production and mothering ability have been the basis for selection. Since 1964, twin ewe and ram lambs have been isolated to form a twinning herd.

They are a large long-bodied sheep, with average fleece weight of 11–12 pounds (58s to 60s count), good lean meat muscling for high cutability, and the special genetic characteristic of twinning.

It is now raised as a closed breed (bringing in no other blood lines) by the Warren Livestock Company. Breeding rams have been sold primarily into California, Idaho, Montana and Colorado, but with some going also into Nebraska and Kansas. A number have gone as far away as the Peruvian Andes to increase the profitability of the Peruvian flocks by multiple lambings.

Warhill ewes and lambs, owned by Warren Livestock Co.

The relatively rare Wiltshire Horn sheep.

WESTPHALIAN

This is another German heavy milking ewe, a large-sized sheep with the ability to rear triplets or quadruplets. It is one of the foreign breeds to watch as it may prove valuable used in crossbreeding to improve milk production.

WILTSHIRE HORN

The Wiltshire Horn sheep is an ancient British hair-sheep breed, once known as the "Western."

In England the rams are used for crossbreeding with smaller breeds of ewes to obtain fat lambs. The lambs inherit the increased fattening quality of the Wiltshire, and can be brought to market on pasture alone. The lambs also inherit the narrow head, which helps in lambing.

This breed has been dwindling rapidly, and also being diluted by cross-breeding. Until recently they were included with the sheep breeds to be preserved by the Rare Breeds Survival Trust of the Royal Agricultural Society in England, dedicated to the protection of breeds that are in danger of extinction.

Pictured here are the first Wiltshire Horn sheep to arrive in Canada, at the ranch of R. A. K. Turner, in Nova Scotia (1972).

WHICH BREED IS BEST?

This is the charted result of a survey conducted by Frank Lessiter, editor of the *National Livestock Producer,* with breeds rated by seven prominent sheepmen, as shown in the October 1975 issue, reprinted by permission.

Breed	Ewe Size	Ram Size	Growth Rate	Feed Efficiency	Muscling	Wool Production
Black-Faced Highland	3.0	3.0	3.0	3.0	3.0	3.0
Cheviot	2.0	1.8	2.3	2.7	2.0	1.7
Columbia	4.8	4.8	3.3	3.5	3.5	4.2
Corriedale	3.3	3.3	3.0	3.0	2.8	4.3
Cotswold	5.0	5.0	3.3	2.7	3.0	3.0
Debouillet	2.5	2.5	2.5	2.0	2.0	5.0
Dorset	3.0	3.3	3.3	3.3	3.3	2.0
Finnish Landrace	2.2	2.2	3.3	3.3	2.6	1.6
Hampshire	4.6	4.6	4.6	4.3	4.6	1.8
Karakul	2.5	2.5	2.5	2.5	2.0	2.0
Large Border Leicester	4.3	4.3	4.0	3.7	3.7	3.8
Lincoln	4.5	4.5	3.3	2.7	3.0	2.8
Merino	2.3	2.3	1.7	2.0	1.7	5.0
Montadale	3.0	3.0	2.7	3.0	2.7	3.0
Oxford	4.0	4.0	4.0	3.3	3.7	2.7
Panama	4.3	4.3	3.7	3.0	3.7	3.7
Rambouillet	4.0	4.0	3.5	3.3	3.6	5.0
Romeldale	3.0	3.0	3.0	3.0	3.0	4.0
Romney	3.0	3.0	3.0	3.0	3.0	3.5
Ryeland	3.0	3.0	3.0	3.0	3.0	3.0
Shropshire	3.5	3.8	3.0	3.0	3.3	3.0
Suffolk	5.0	5.0	4.8	4.8	5.0	1.0
Southdown	1.2	1.2	1.2	2.5	3.0	1.8
Targhee	4.3	4.3	3.8	4.0	3.5	4.5
Tunis	3.0	3.0	3.0	3.0	3.0	2.5

*Denotes tie.

The breeds were compared as to their potential value to a commercial sheepman — with a flock of 250 grade ewes — who was willing to crossbreed. The breeds were scored by seven sheep experts as follows: 5.0—Excellent; 4.0—Good; 3.0—Average; 2.0—Unsatisfactory; and 1.0—Poor.

Wool Grade	Out of Season Breeding	Ease of Lambing	Milking Ability	Longevity	Hardiness	Overall Breed Ranking
2.0	2.5	3.0	2.5	3.5	2.5	22nd
3.0	2.0	4.0	2.0	3.0	3.8	23rd
4.0	2.4	3.3	3.5	3.0	3.4	3rd
4.0	2.7	3.7	3.7	4.0	3.7	7th
2.0	2.3	2.3	2.3	2.7	1.7	19th
5.0	3.5	4.0	3.0	4.0	4.0	9th
3.0	4.3	4.0	4.0	3.0	3.0	10th
2.0	4.0	5.0	4.4	2.7	3.0	15th
2.7	3.0	2.7	4.0	2.3	3.0	5th
2.0	2.5	2.5	2.5	3.0	2.0	24th
3.3	3.0	3.0	3.0	3.0	3.0	6th
2.0	2.3	2.3	2.3	2.7	2.3	21st
4.7	3.0	3.0	2.3	3.7	4.3	16th*
3.0	2.7	3.0	3.0	3.0	3.0	20th
2.7	2.3	3.0	2.7	2.7	3.0	11th
3.0	3.0	3.0	3.0	3.0	3.0	8th
5.0	4.2	3.8	3.4	4.8	4.8	1st
3.5	3.0	3.0	3.0	3.0	3.0	12th
3.0	3.0	3.0	3.0	3.0	3.0	14th
3.0	3.0	3.0	3.0	3.0	3.0	16th*
3.0	2.3	3.3	3.0	3.3	3.0	13th
2.0	3.3	4.0	4.4	1.6	2.5	4th
2.3	2.0	2.7	1.7	3.3	3.0	25th
4.7	3.0	4.3	3.8	4.0	4.3	2nd
2.5	3.0	3.5	3.5	3.0	3.0	16th*

43

Breed	Heavy Weight When Fat (Lambs)	Grazing Ability	Heat Toler- ance	Prolif- icacy	Mother- ing Ability	Temper- ament
Black-Faced Highland	3.8	5.0	3.0	3.0	3.5	3.0
Cheviot	1.5	5.0	3.5	1.0	5.0	1.0
Columbia	5.0	4.2	4.5	3.5	4.0	3.8
Corriedale	3.5	4.0	4.5	3.0	3.2	3.6
Cotswold	5.0	1.5	2.5	3.5	2.5	3.0
Debouillet	3.0	4.5	5.0	3.0	3.0	3.0
Dorset	3.0	3.2	4.0	3.5	4.5	4.0
Finnsheep	3.5	3.0	3.0	5.0	4.5	2.5
Hampshire	4.5	3.8	4.0	3.2	4.5	4.0
Karakul	3.0	4.0	4.5	2.0	2.5	2.0
Border Leicester	4.8	2.8	3.0	3.5	2.5	3.0
Lincoln	5.0	2.0	3.0	3.5	3.0	4.0
Merino	1.5	5.0	5.0	1.0	2.0	1.0
Montadale	3.0	4.5	3.5	3.5	4.0	1.0
Oxford	4.0	3.0	4.0	4.0	4.0	4.0
Panama	5.0	4.2	4.5	3.2	3.5	3.5
Rambouillet	4.2	4.5	5.0	3.0	3.2	3.0
Romeldale	3.0	3.5	4.0	2.5	3.8	3.5
Romney	3.5	4.0	3.5	3.0	4.0	4.8
Ryeland	3.0	3.5	3.0	2.5	3.0	4.0
Shropshire	3.0	3.5	3.5	3.5	4.5	4.0
Suffolk	5.0	4.0	4.0	3.5	4.5	2.8
Southdown	1.5	1.0	1.0	2.5	4.5	5.0
Targhee	4.5	4.5	4.8	3.2	3.5	3.5
Tunis	3.0	3.5	5.0	4.0	4.0	3.0

These additional traits were rated by Robert M. Jordan, professor of Animal Science, University of Minnesota, and some of these were printed in *Shepherd* magazine in March, 1976.

Some traits are of importance in one sex and not visibly important in the other, except as the trait is passed on to the offspring. Examples are prolificacy and mothering ability of the female, and heat tolerance, which is primarily of interest in the male by way of sterility in very hot weather.

The ratings of many of these traits are not permanently fixed; they can still be improved by breeding and selection. This is most noticeable in prolificacy, for if you retain only twin ewes for breeding purposes, and use rams that are also twins, you can increase your lambing percentages above what might be expected of a less prolific breed.

Management and nutrition also play a large role in the exploiting of the potentially profitable traits in *any* of the breeds.

CHAPTER THREE

FENCES AND PASTURES

More than 90 million acres in this country cannot be used efficiently for anything other than growing grass, and some parts of all farms are probably less suitable for crops than for pasture.

Sheep are more efficient than cattle in converting grass to meat. They have their lambs in the spring, so the lambs grow to market age on the abundance of summer grass and can be sold about the time the pasture gives out in late summer or early fall. This means that you do not need to carry the meat animals through the winter on hay and grain, as you would beef animals.

They also distribute their droppings evenly and widely, improving the pasture rather than harming it. Australian research has determined that a cow, on the other hand, blankets an area of 7.5 percent of an acre each year with impenetrable dung pads. The rank growth that comes up around each pad further reduces the amount of good grass, so that the total loss of production may amount to as much as ⅕ acre per cow, each year.

PASTURES

STOCKING RATE

How many sheep can be kept per acre? Opinions differ. The fewer breeding ewes grazing during the winter, the less supplemental feed they will need at that time. Sheep do not do as well when pasture is overstocked, and the older ewes suffer the most. Their poor teeth make it harder for them to chomp on overgrazed pasture, and with short grass they obtain less feed per bite.

Some farms estimate four sheep to an acre of good pasture, with hay and some grain in the winter, and one or two sheep to an acre of poor pasture, again with supplemental food in winter. So you should take a good look at the condition of your acreage. Better to keep too few the first year, and see how the pasturage holds up.

A number of factors are involved in deciding how many sheep your pasture will support. These include: Type of soil (rock, sand, clay, etc.); plant species (grass, weeds, clover, etc.); rainfall or irrigation; climate; fertility of soil; lay of the land (hill, slope, marsh); whether lambs or ewes with lambs, or dry ewes; and whether pasture is rotated.

ROTATION

Some plants, such as alfalfa, cannot stand continuous grazing, but can stand hard use over shorter periods of time. Alternating pastures gives the plant growth a chance to recover, and gets more actual growth out of the same amount of space. Most grasses grow better when given a periodic rest from grazing.

On large farms, it has been determined that 100 sheep will do better on twenty-five acres that is divided than on thirty acres with no division. This holds true on smaller farms, too.

Sheep by nature would prefer not to graze continually in the same place. They like a fresh pasture that has not been walked on. If with ample pasture they seem choosy and walk around with their noses to the grass but not eating as much or chewing their cuds as much, it is time for a pasture change. They need not even be moved to one where the grass is much higher. It will have "fresh" grass, and they will eat better.

Perennial grass stores food in its roots after it has made the season's main growth. The grass uses these reserves to survive while dormant, to make the first spring growth, and to start new growth after its leaves are closely grazed. Its ability to recover quickly after grazing makes grass valuable for forage production, but can deceive us into thinking that leaves can be repeatedly removed without injury. If they are, the plant keeps drawing on food stored in the roots, to grow new leaves, until the supply is exhausted and the grass dies.

Many grasses will not reach their maximum vigor and growth when more than half of their leaf surface is removed. This weakened grass does not make efficient use of soil moisture and nutrients and does not provide the maximum livestock feed.

During the droughts of the 1930s and 1950s, wind erosion occurred largely on land that had little or no plant cover because of cultivation or too close grazing. On farms, grass pasture not only feeds the sheep, but protects the soil from erosion by wind or water—if you rotate your pastures and prevent over-grazing.

When fencing an area, divide it into at least two pastures, or three if it is large. This gives the grass a rest and helps to control parasites. Since wire fencing is expensive, and time-consuming to install, this is not always feasible. Cross fences need not be as heavy, or as high as the outer fence, nor need they be dog-proof.

When rotating pastures, don't let the grass get too tall before you turn the sheep in, for they will trample more of the grass and will not eat it as well as they would shorter grass. From 4 inches to 6 inches is ideal.

SHEEP AND GOAT PASTURE

If the pasture was cleared once, but has been without livestock for a while and is partly grown up with blackberries, Scotch broom, small saplings and brush, you can run goats with the sheep to clear the brush. While sheep like to eat from 4 inches to 8 inches from the ground, goats like to eat from about 10 inches to as high as they can reach, and are great brush and bramble clearers. Sheep are grazers, goats are browsers.

ORCHARDS

An orchard is one of the favorite sheep pastures on a small farm. The sheep can make use of the shade in the summer, and if a little care is exercised to prevent them from getting too many windfall apples and other fruit, they can make good nutritional use of fallen fruit. They should not be turned suddenly into an orchard with a lot of fruit on the ground. However, if they are there all the time, they are accustomed to the fruit in their diet. The extra fruit should be picked up and taken away, so they don't get too much. This fruit will keep long enough to be added to their diet after the fallen fruit in the orchard is gone. We put away windfall apples and dole them out, a few at a time, far into the winter. Try to save some apples until lambing time, so they can be given as treats to the ewes when they are in the lambing pens.

You will find an occasional sheep with goat-like habits, standing on its hind legs and nibbling the branches and leaves on the fruit trees. Usually sheep don't do that, but they will chew on the bark of the trunks, and can do a lot of damage if you do not protect the trees. The trunks of larger trees can be wrapped with several layers of chicken wire, or once around with rabbit wire. Old burlap feed bags can be used, and fastened with wire or baling cord. A temporary solution is to make "manure tea" from sheep droppings, and paint that on the trunks. This must be repeated after every rain.

Any small or newly planted trees will need the protection of a rigid fence. A board fence around small trees provides secure protection for the tender tree, and is easy for people to climb over to prune or to pick fruit. Avoid having the fence so close to the tree that sheep will brace their feet on it to reach the lower branches.

If you buy an old farm, the fences and barn buildings will probably need repairs. The buildings can be repaired after you get the sheep, but the fences should be done before. Sheep quickly learn to jump sagging fences. One sheep loose in a neighborhood can be quite a problem, and a sheep in your garden is a disaster.

If you wait until they have the habit, they may still do it after the fence is repaired. One jumper can set a bad example and should be sold, or slowed down by temporary "clogging" until retrained. Fasten a piece of wood to one front ankle with a strap. It will get in the way just enough to prevent the sheep from jumping.

POSTS

The life of any fence depends a lot on how hard the sheep worry it, and how long the posts, especially the end posts, hold up. They should be massive and solid, for if they start rotting and the sheep rub on them, or put their heads through the fence and strain to reach greener grass on the other side — there goes your fence.

Painting posts can make a fence more attractive, but not more durable. Posts tapered at the top to drain off rain and snow sound reasonable, but tests show no improvement in the life of the posts. The only reliable way to prolong the posts' usefulness is with wood preservatives.

ESTIMATED LIFE OF UNTREATED
WOOD POSTS
(Round, 5 to 6 inches in size)

Over 15 Years	7 to 15 Years	3 to 7 Years	
Black Locust	Cedar	Ash	Honey Locust
Osaga-orange	Red Cedar	Aspen	Maple
	Red Mulberry	Balsam Fir	Pine
	Redwood	Beech	Red Oak
	Sassafras	Box Elder	Spruce
	White Oak	Butternut	Sycamore
		Douglas Fir	Tamarack
		Hemlock	Willow
		Hickory	Yellow Poplar

Split posts, which have more "heartwood," will last longer than the time listed, and larger sized posts also last longer. In addition, the life of the post is doubled or even tripled when it is properly treated.

If untreated posts are set in concrete, or thinly coated with concrete, water can get in cracks between the post and the concrete, and moisture will be held in, so the posts will rot faster. Some farmers pile stones around untreated posts to keep back the weeds and promote air movement around the posts, but the stones hold moisture at the ground line, and encourage decay there.

Do not use paint containing lead on sheep equipment or on parts of buildings accessible to sheep. Poisoning may result when animals constantly lick or chew objects covered with paint containing lead.

PRESERVATIVES

Coal-tar creosote is the best preservative. Since heat and special equipment are required, this is a difficult process for the home owner. There are many other ways of treating wooden posts to extend their usefulness, but only two of them are practical for the small farm, cold soaking on seasoned posts, and end soaking on green posts, which is not as satisfactory.

END SOAKING

If you need a fence right now and don't have time to cut your posts ahead and season them before treatment, end soaking is the method to use. Cut round posts and leave the bark on the posts. Then use a 15 percent to 20 percent solution of zinc chloride, or chromatized (chromated) zinc chloride in water, (which is non-toxic to humans or sheep). Allow about 5 pounds (or about a half-gallon) of the solution for each cubic foot of post to be treated. The chromated zinc chloride is sold in a granular form that is easy to use, and is less subject to leaching from the posts than the plain zinc chloride. Often it is difficult to find a source for this chemical. If you can't find it, you usually can buy its two ingredients from a chemical company, and mix them. Use 80 percent zinc chloride and 20 percent sodium bichromate. For the solution mix a 20 percent chemical, 80 percent water solution (each by weight). Thus 100 gallons of water weigh about 830 pounds and would require 166 pounds of the zinc mixture.

Stand the posts bottom down in a tub or drum of this solution until they absorb about three-fourths of it, which takes from three to ten days. Then stand them on their tops, and let them absorb the rest. They should be seasoned for about a month before using, to allow the treated wood to dry.

Either green or seasoned posts may be soaked, covered completely and steeped, in a 5 percent solution of this same zinc chloride, for from one to two weeks, but this is not as effective as a pentachlorophenal solution treatment.

Oil drum used for fence post treatment.

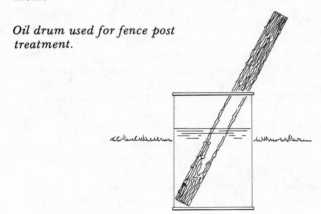

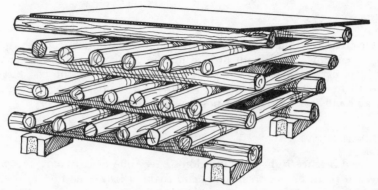

Proper storage of fence posts, for drying. Note "roof" on top of pile, to keep off rain.

COLD SOAKING

The best absorption and penetration are obtained by first seasoning the posts. This lets the sap dry out to make room for the preservative. Peeled posts should be open piled, so that the air can circulate around each one, and the bottom of the pile should be at least a foot above the ground. The best place for piling would be an exposed area on well-drained ground.

While posts cut in the spring will peel more easily, posts cut in the fall will have a chance to dry more slowly, which prevents some cracking and checking. This is more important with oak posts than with wood from cone-bearing trees.

The seasoning of posts adds little to their life *unless* they are also treated with preservatives.

Cold soaking of seasoned posts. This consists of soaking in a solution of *pentacholorophenol* (called *penta* for short), or in *copper naphthenate,* diluted with either fuel oil or diesel oil. Wear rubber gloves, for these irritate the skin. Penta can be purchased in a concentrated solution, or in a ready-to-use solution (more expensive) or in flakes. If you use penta in dry flakes, wear goggles and a dust mask when mixing, to avoid irritating the eyes and throat. You can see why the concentrated penta solution is the most popular. It is made in several strengths, calling for dilution by mixing with two to twelve or more parts of oil, to make a 5 percent solution. The label should specify strength and amount of dilution. Fifty gallons of a mixture of oil and penta will treat fifty posts of 6-inch diameter and 6-foot length. For convenience, they can be soaked in upright drums, soaking the bottom portion longer, for the tops are less subject to decay. (This is also a good treatment for seasoned boards for use in a board fence.)

After removal from the solution, posts should be stacked so that the excess solution dripping from each post will be absorbed by the post beneath it. Wear gloves when handling them until they are dry. Penta is highly toxic and can be absorbed through the skin and by way of the lungs. Animals should be kept away from newly treated wood. Do not treat wood for feed troughs. Store penta carefully, away from the reach of children, for there is no antidote for penta poisoning.

STEEL POSTS

The use of steel posts can save a lot of labor in digging and tamping dirt around posts. It is even more of a time- and labor-saver when you are fencing through a wooded area where there are roots and stumps. Driving a slender metal post is easier than digging a hole.

Steel posts are driven into the ground with the use of a 16-pound "sleeve" that fits over the top of the post. While it is suggested you pound it with a sledge or by hand, the latter — raising up and pounding down over the post — is the simple way. There are also heavier spring-loaded post driver sleeves, with handles on the sides.

Galvanized steel posts have the longest life, followed by posts brush-painted with metallic zinc. Posts that are dip-painted with lead and oil paint can start to rust within five to seven years.

Some posts are sold with five clips included in the price of each post. If not, order the clips at the same time as the posts, so you get the right style.

Old iron bed-rails where *springs* were fastened, often found in a salvage yard, make good posts. Some already have holes in them where the springs were fastened. If not, you may want to drill some holes to fasten the fence.

Discarded pipe from machine or repair shops is sometimes much cheaper than steel posts. It should be at least 1¾ inches in diameter, and heavier than that for corner posts. End posts of 6-inch or 8-inch pipe can be filled with concrete.

Concrete can be poured around the steel post, with the "form" dug in the ground. The hole-form should extend 3½ feet into the ground, about 18 inches square at the top, and 20 inches square at the bottom. Posts should be maintained about 1 inch out of plumb, away from the direction the fence will be stretched, while pouring the cement. The braces for the steel

Two styles of "sleeves" for driving steel posts.

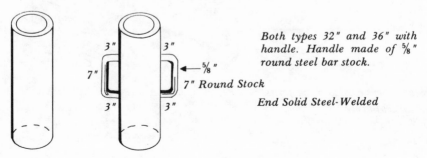

Both types 32" and 36" with handle. Handle made of ⅝" round steel bar stock.

7" Round Stock

End Solid Steel-Welded

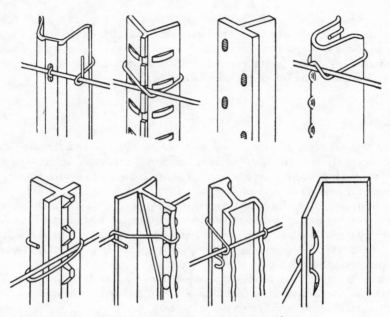

Steel fence posts and methods of fastening wire to them.

corner posts should be attached to the corner posts before positioning the ends that will be sunk in concrete. Determine location of concrete pier, and dig form about 20 inches square, and 18 inches deep. The brace will enter the concrete about 6 inches below the ground, and extend at least 6 inches into the concrete.

DEPTH OF SETTING

Line posts are usually set $2\frac{1}{2}$ feet in the ground. End, corner and brace posts are set $3\frac{1}{2}$ feet deep, and gate posts are set 4 feet deep.

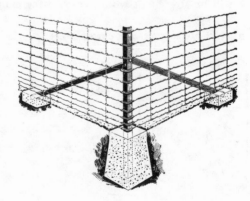

Steel corner and brace posts set in concrete.

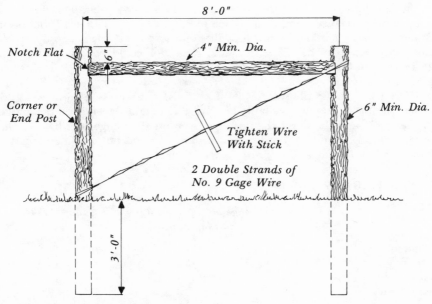

Corner or end post assembly.

Hand augers or posthole diggers are easier to use than shovels, and there's less dirt to put back into the hole. In heavy soil or clay, oil the posthole digger so the dirt does not stick to it. Keep a bucket of waste crankcase oil where you are digging, and dip the digger into it frequently. Posts should have the dirt tamped tightly around them, and should be aligned with the rest of the fence posts while being tamped.

The longest lasting posts are purchased ones, pressure treated with coal-tar creosote, with an average life of at least thirty years, but initial costs are high, and they are not available everywhere. A fence is no stronger than its end posts and braces. The brace wire has its ends spliced together, and is tightened by twisting it with a strong stick or rod. Leave the stick in place so you can adjust it as necessary.

FENCING

WOVEN WIRE FENCING

Woven wire fencing is the most practical fence for sheep, at least for the bottom half of the fence. Wire "stock fence" comes in different weights and styles. The five most common style numbers are 1155, 1047, 939, 832 and 726. The last two numbers denote the height of the wire in inches, and the first one or two numbers indicate the number of horizontal (line) wires in the fencing. There are three other styles, not seen as often: 949, 845, and 635. All the heights come in a choice of stays that are either 12 inches or 6 inches apart. While 6-inch stays will stop more dogs, the sheep don't get their heads stuck in the 12-inch stays. A small dog could possibly go through 12-inch stays, but a dog that small could probably get through lots of other places.

Woven wire is usually sold in 330-foot rolls, in a choice of four weights, although not all may be available locally. The weight depends on the gauge of the wires, and the lower the number given for the gauge, the heavier the wire: 9, 11, 12½ and 14½ are standard. The top and bottom line wires are heavier than the stay and intermediate line wires.

FENCE WEIGHTS

Weight	Gauge of Top and Bottom Line Wires	Gauge of Filler and Stay Wires
Light	11	14½
Medium	10	12½
Heavy	9	11
Extra heavy	9	9

Woven wire is coated with aluminum or zinc (galvanized). The zinc coating can be of several different thicknesses, indicated by government class numbers: 1, 2, or 3, class 3 being the heaviest coating.

The lightweight fencing is suitable only for cross fences, will not last as long as a heavier wire, and is easier for the sheep to wear down. The heavy and extra-heavy wires are hard to work with, especially if they are in 6-inch

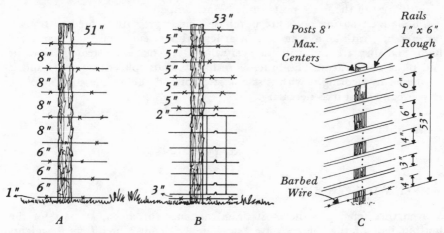

Three types of fencing for sheep. A is all barbed wire, B is a combination of fencing and barbed wire, and C is a board fence with barbed wire at the base.

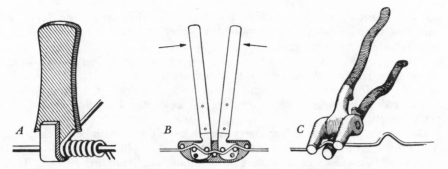

Useful fencing tools include (A) splicing tool, (B) double-crimp tool, and (C) single-crimp tool.

stays, or very high. Most people settle for the medium weight, so there is more demand for it, and it is easier to find.

A height of 26 inches or 32 inches can be handled with relative ease, and you build up the height with barbed wires. Place the first wire 1 inch above the top of the woven wire, for if the woven wire sags even a little, that is where the sheep will get their heads through. Also, string barbed wire about 1 inch below the bottom of the woven wire, to discourage dogs. When you stretch woven wire over uneven ground, there are places where it doesn't touch the ground. You can make up this difference with barbed wire along the bottom, parallel to the contour of the ground. A third wire half-way between the top and bottom of the woven wire will help reinforce the fence by discouraging the sheep from rubbing on it, or putting their heads through.

Putting up fencing can be dangerous, but you can minimize the risk of injury by wearing heavy leather gloves, high boots, and tough clothing.

STRETCHING WOVEN WIRE

Woven wire can be stretched only from one anchored post to another, but an anchored post can be a gate, corner, end or just a braced-line post — any but the unanchored line posts. For the sake of simplicity, we will assume that the fence is being stretched from one wooden end post to another.

Set the roll of fencing on end, about 1 rod (16½ feet) from the end post. To prevent staples from being pulled by the sheep pressing against the fence, put the wire on the livestock side of the posts. Where appearance is important, you may want to put it on the outside.

Unroll enough wire to make a wrap around the post, and fasten back onto the fence. Remove two or three stay (vertical) wires, depending on the circumference of the post, and position the next stay wire along the edge of the post. Starting with the center line wire, wrap each wire around the end post and wind it back around the line wire, five wraps.

Unroll the fencing, keeping the bottom wire close to each post. Before stretching the wire, tie it loosely to the posts with baling twine, or prop it up with temporary stakes. Place the clamp so that when the fence is stretched the clamp will pass beyond the other end post, enabling you to staple the fencing to the post.

You will only need a "single" stretcher on 26-inch or 32-inch woven wire. Some stretchers must be anchored to a dummy post or a tree.

To be avoided are the rope-and-pulley types that don't give enough control to avoid accidents, and require a great deal of pulling and tugging to get a fence tight.

Before you start stretching the woven wire, note the shape of the tension curves (the little crimps in the line wires). The fence is stretched properly when these tension curves are about half straightened out. This provides for expansion and contraction with changes of temperature, and also leaves tension in the fence, which makes for a tighter fence.

As a safety precaution when stretching, stand on the opposite side of the post from the stretcher so if something gives, you'll be out of the way.

Fasten the fence to the posts on the ridges and in the low places first, fastening the line wires one at a time, starting at the top or bottom, whichever is the tighter.

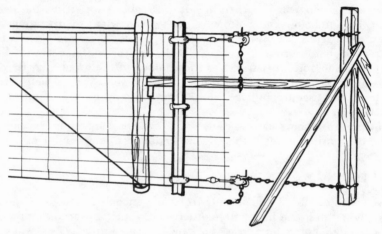

Fence stretcher in position. Note braced "dummy" post at right.

To counteract the tendency of the tightly stretched fencing in a low place pulling up the post, you can bury a "dead man," an old log, under your fence line at that place. Secure the bottom of the woven wire to the buried log, and it will do the work of a couple of earth-anchors, at no cost. A large boulder can be used in place of the log.

When fastening the woven wire to the end post, secure each line wire with two staples angled in opposite directions to prevent slippage. Measure the circumference of this post. Cut the fencing, allowing that many inches plus at least 6 inches more, for five wraps around the line wires. Remove a couple of stay wires, wrap the fence around the end post and secure it by

A neat job of splicing woven wire. (USDA photo)

wrapping the ends of the line wires back around the corresponding line wires of the fence. If the wire continues around a corner post, there is no need to cut it. Just fasten it and continue around.

Woven wire should be stretched and attached in sections running from one anchor post to the next.

BARBED WIRE

Barbed wire is usually available in both 12½ and 14 gauge, with the lighter (14) used mainly for temporary fences. There are also two-point barbs and four-point barbs, but the number of barbs is more important if you are running cattle, where the complete fence would be made of barbed wire. It also has the same "class" numbers to denote the thickness of the zinc or aluminum coating.

These instructions apply specifically to stretching barbed wire above woven wire. In such a fence, barbed wire is attached after woven wire is in place.

Before unrolling barbed wire, attach one end to the gatepost or end post. Two people can unroll a reel of wire by placing a rod through the center of the reel and letting it unroll as they walk down the fence line. (Nasco Catalog has a dandy Barb Dollie that makes it a one-man job.) Two or more reels can be unrolled by placing them on an end-gate rod in the back of a truck or wagon.

The handy "Barb Dollie," in position for pulling off barbed wire. Handle position can be reversed for easier moving of wire spool.

The rope-and-pulley stretcher, which I wouldn't recommend for woven wire, can be used to stretch barbed wire. But you can stretch it tight enough with a crowbar or a claw hammer so that any slack will be taken up when you staple it to the posts.

If you buy a farm with an existing barbed wire fence for cattle, the practical thing to do would be to add additional wires to make it more dog-tight and usable for sheep.

ELECTRIC FENCE

Our own experience with electric fence was not too encouraging, but we were trying it for sheep and goats both, on wooded pasture. After hanging chains on the goats' collars so they wouldn't go over, and wire antennae on their horns so they wouldn't go under, and still having our problems, we just gave up.

On a cleared field, sheep only, it probably works out all right. Sheep will have to be trained to respect the fence by nose-contact, for their wool insulates their bodies. Electric fences sometimes have metal strips hanging from them, which will be sniffed by the sheep. There are electric barbed wires available that are more likely to penetrate the wool of sheep or the hair of goats.

A new electric fence called *Flexinet* is in use in Britain. It is woven wire electric fencing, designed for sheep. Plastic posts are permanently attached to the woven wire that is made of stainless steel coated with plastic. The manufacturer says that 100 yards of fence can be set up easily in twenty minutes. The manufacturing company is behind in orders for it, but this fence is expected to be marketed here, in time.

When considering electric fences, check with your county extension agent, for ordinances differ from one area to another regarding the installation of electric fences, and the legal types of chargers and control boxes. Fiber glass posts are available that are especially suited to electric fence use.

WOOD FENCES

Board, slab or plank fences are good rigid fences for corrals, and attractive fences around a farmhouse. For corrals, the boards should be on the inside of the fence, so sheep will not loosen them. Posts would need to be closer together than for pasture fences, and the planks well spiked or bolted to them.

DOG-TIGHT FENCES

While sheep on open range and in less settled areas still have terrible problems with coyotes, in the populated places and small farm communities, the domestic dog does the damage. The U.S. dog population is estimated at 22 million, which is about three dogs for every two sheep.

A dog need not be wild or vicious or even brave to chase a sheep. They are only following their natural impulse to chase whatever runs. Unfortunately, sheep startle and run at the slightest disturbance.

In addition to good fences, dogs can be discouraged by:

1. Bells on some of your sheep. You can hear the bells if the sheep are being chased. High-frequency bells have also been tried with some success, as dogs find the sound unpleasant.
2. Having sheep in an open field, in sight of your house.
3. Using coyote snares, if your problem is really coyotes. These are listed in the Sources chapter.
4. Having a dog of your own, who will make a commotion if another dog comes by.
5. Having a gun if necessary. Even a pellet gun can drive off an attacking dog.

Coyote snare being placed under fence. (West Texas Livestock Weekly)

A dog too small to kill or maim sheep can still cause death by heart failure in old ewes and abortion in pregnant ewes. Broken legs, too, often result when sheep are chased.

For a dog-tight corral fence, the 12-inch stays are not suitable, and ones even smaller than the 6-inch would be best. We have used the "non-climbable horse fence" wire, and found that it is very satisfactory for a small area, but so rigid as to be hard to stretch unless you have a straight pull. Whatever you use should be extended to at least 5½ or 6 feet by barbed wire, to be a foolproof corral. Don't neglect the one strand of barbed wire at the ground level, on the outside of the posts, to discourage digging.

DOG LAWS

Your county agricultural agent should be able to tell you the county's dog laws, or better yet, to give you a copy of the county or state laws.

Sheep owners should know the law, and work for more adequate dog control legislation if it is needed. These laws should permit the elimination of any trespassing dog that is molesting livestock. They should require

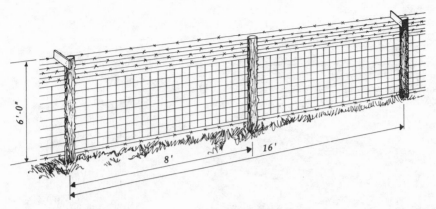

This combination woven fence and barbed wire is built to protect sheep from dogs and coyotes.

payment by the owner of the offending dog for both damage and deaths to livestock.

The laws should require prompt action by enforcement agencies, and should require payment by the county or state to the livestock owner for losses incurred to his sheep (or other livestock) from any unidentified dog, or dogs, not apprehended by the proper officials. When the laws do require county or state payments, you should follow up on this. While the amount paid is seldom adequate payment for the loss, enough claims of damages will make the county officials more strict with their dog laws.

Many states and counties already have these laws on their books, but the sheep owners do not always know the law.

GATES

Any gate the sheep will be using regularly should be a wide one. Narrow gates and narrow doorways are dangerous when ewes are pregnant, for being crowded and bumped can cause abortions.

It is far easier to lead sheep through a gate than to drive them. If they are in the habit of getting a bit of grain now and then, even at times of the year when it is not completely necessary, you have a way of controlling them. Just rattle the grain in a pan and they will follow you.

There are gate-hardware kits in supply catalogs, having all the materials for the gate with the exception of about five boards.

With wide gates, you will need a heavy gate post, and if it is a tall one, you can suspend the latch end of the gate to the post with a cable so the gate will not sag. If you use a turnbuckle in the cable, it can then be tightened as necessary to avoid sagging.

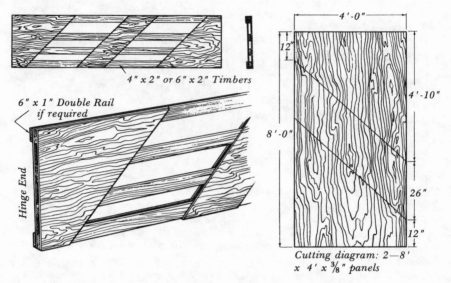

4" x 2" or 6" x 2" Timbers

6" x 1" Double Rail
if required

Hinge End

4'-0"

12"

4'-10"

8'-0"

26"

12"

Cutting diagram: 2—8'
x 4' x ⅜" panels

Two 8'x4'x⅜" sheets of waterproof plywood are cut as shown at right to make this tough, long-lived gate. Rust-resistant nails or screws are used to fasten plywood to 2"x4" or 2"x6" timbers.

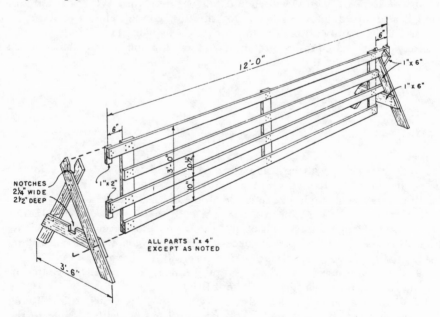

12'-0"

6"

1"x 6"

1"x 6"

6"

1"x 2"

3'-0"

3'-0"

NOTCHES
2¼" WIDE
2½" DEEP

ALL PARTS 1"x 4"
EXCEPT AS NOTED

3'-6"

This portable sheep fence is easy to move and handy wherever a temporary fence is needed. (Utah State University)

With heavy wide gates, the use of an old wheel from a discarded wheel-barrow to support the latch end will make the gate roll open and closed easily, and the gate will not sag from its own weight.

The illustration shows a plywood gate designed by Forest Industries of Canada, which uses two sheets of plywood cut diagonally into three pieces, to make a gate any width from 8 feet to 16 feet. It should be assembled with rust-resistant nails or $\frac{3}{8}$-inch diameter galvanized bolts. The exterior plywood gate is not subject to joint shrinkage and transverse stresses, and may be expected to last twenty years or more if painted with wood preservatives.

For gates in cross fences, it is a good idea to anticipate "forward creep grazing" and make allowance for adjustable openings where the lambs are allowed to go into new pasture ahead of the ewes.

There are several advantages in placing gates between pastures in the corner where the cross fence joins the outer fence. A braced-line post can serve as one of the fence posts. And it is easier to get the sheep through a gate that is in a corner rather than in the middle of a fence. When you need to pen the sheep for shearing or hoof trimming, or catching them for any reason, you can set up a temporary corral with just two movable fence sections, and drive or lead the sheep through the corner gate into this enclosure. With one movable fence section you can set up a chute, with the outer fence as one side of it, for loading sheep into a vehicle.

Inexpensive hinges for narrow people-gates can be rectangles cut from old auto tires. They are flexible and easily replaced when they do wear out. The convex rubber, mounted properly, also keeps the gate closed.

CHAPTER FOUR

FEEDS AND FEEDING

Raising sheep is an efficient way to convert grass into food and clothing for humans, but pasture alone is seldom adequate for sheep twelve months of the year, making the feeding of supplements quite necessary.

Poor feeding of ewes results in reduced fertility, reduced growth of young, poor nursing ability of ewes, fewer multiple births, decreased wool production, and higher incidence of pregnancy disease.

An undernourished ewe may also lamb a few days early, and if the lamb has not reached its full size before birth, it has less chance of survival. An undersized lamb, born outdoors in bad weather, suffers considerably more loss of body heat because of its smaller size.

At feeding is a good time to check on your sheep, feel their udders when close to lambing, and note eating habits which greatly reflect their state of health. Count the sheep, particularly if you have any wooded pasture where one could get snarled up or be down on its back and need help.

FEEDING

GRAIN FEEDING PERIODS

1. Seventeen days before turning ram in (flushing before breeding). Give $\frac{1}{5}$ to $\frac{1}{2}$ lb. per ewe.
2. Up to three to four weeks after mating, give same amount, tapering off gradually. This may prevent resorbtion of the fertilized ova.
3. Last four weeks of pregnancy, when ewe should be on rising plane of nutrition to prevent pregnancy disease, give same amount.
4. For six weeks of lactation, same amount, then tapering off gradually as the lambs eat more grain and hay.

If building a barn or shelter, plan for a feed room, or some safe place for storing grain where sheep can't break in and get it. This can be a closed

A pan of grain makes the author a popular person with her sheep. (Josef Scaylea)

room, or alcove with gate. Put feed in a garbage can with the lid wired down, or a spring-clamp that will keep the lid on, even if it is tipped over. This has the added advantage of being rodent-proof. Any sudden large amount of grain can paralyze the digestive system of the sheep, and cause death from impacted rumen or bloat. In an animal with four stomachs, "acute indigestion" is not a minor illness.

REGULAR FEEDING (TIME AND AMOUNT)

Measure the quantity of grain given each day, by using the same container, or number of containers, for each feeding. Sheep do not thrive as well when the size of their portions fluctuates. If fed in the evening, it should be at least an hour before dark, for sheep are not like cattle and horses; they do not eat well in the dark and should have time to eat their food before nightfall.

When given regular feedings at an expected time, they are less apt to bolt their feed and choke. Too much variation in feeding time is hard on their stomachs and their entire systems.

GRAIN

Whole grains, with the exception of barley which can lead to metabolic disturbances in pregnant ewes, are better for sheep than crushed grain. Rolled oats, for instance, has so much powdery substance that it can cause excessive sneezing, leading to prolapse in heavy pregnant ewes, and breathing problems in lambs. Unprocessed corn and wheat still contain the valuable germ rich in Vitamin E that ewes need to help protect their lambs from white muscle disease.

Whole grains promote a more healthy rumen. Pelleted feed causes the papillae of the rumen to lump together and become inflamed. This traps debris and causes more inflammation. Whole grain, on the other hand, promotes a healthy rumen wall, where the feed gives better conversion to growth.

FEED CHANGES

A sheep's stomach can adjust to a great variety of feed, providing changes are made *gradually*. A sudden change of ration, such as sudden access to excess food, can cause death. The rumen has a mixed bacterial content with the ability to adapt to the nature of the diet. Sheep who are fed only grain will not be able to adapt to a sudden change to hay, for the rumen will be so geared to the handling of concentrated starch and protein that the bacteria which digest cellulose will be present in too small a number to

An old hot water tank, cut in half lengthwise, makes a good outdoor feeder for grain. (Charles R. Pearson)

Suffolk sheep find this hay and grain feeder a good gathering place. Drawings show how to construct it. Board C can be made of 1" x 10" for large ewes, or from 1" x 8" for smaller breeds. For use with lambs or small breeds, the dividers B can be made from 1" x 6" instead of 1" x 2". (Photo from National Suffolk Sheep Association. Drawing from Midwest Plan Service)

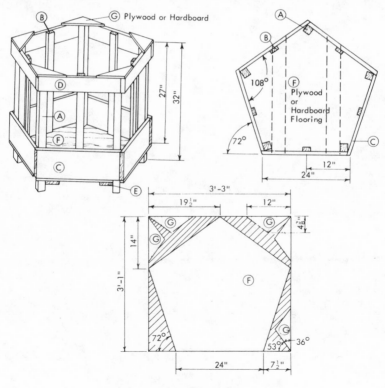

function, and the sheep will go off its feed and suffer. A gradual change from grain to hay gives those cellulose-handling flora a chance to multiply. A disturbance of the rumen by abrupt change of diet will leave the sheep open to infections and disease, by interfering with the synthesis of A and B vitamins, vitamin A in particular being the anti-infection vitamin.

HAY

One reason alfalfa hay is such a superior feed for sheep is its content of nine vitamins, especially Vitamin A that is so lacking in winter pasture grass. The greener the hay, the higher the vitamin content. It is also high in calcium, magnesium, phosphorus, iron, and potassium. Protein content is from 12 percent to 20 percent depending on what stage it is cut (highest protein when cut in the bud) and on its subsequent storage. Alfalfa got its name from an Arabic word meaning "best fodder," which is most appropriate.

Hay should be stored in the darkest part of the barn to preserve its Vitamin A, which is depleted by exposure to sunlight. Careful storage is necessary to avoid weather damage and nutrient loss. Exposure of hay to rain can not only leach out its minerals, but can result in mouldy hay, one cause of abortion in ewes.

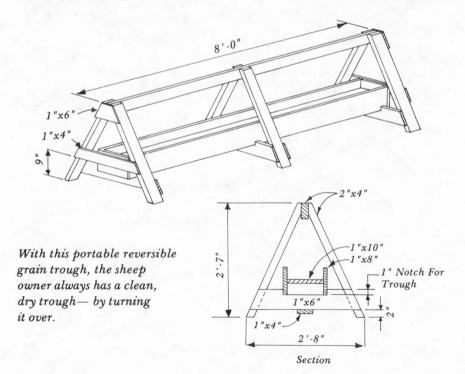

With this portable reversible grain trough, the sheep owner always has a clean, dry trough— by turning it over.

Section

The lower the hay quality, the more of it you will have to feed. Lots of heavy stems in the hay will mean more that the sheep will not eat. A certain amount of hay is always discarded, some pulled out onto the ground and wasted (pile this in your garden twice a year) and some uneaten stems (save these clean ones out of the feed rack for clean bedding for lamb pens).

If you buy two different kinds of hay, save the best for the pregnant ewes. Late in pregnancy, hay *must* be of the high quality, as the growth of the lamb will crowd the ewe's stomach and leave little room for bulky low-nutritive feed.

EXTRAS

Windfall apples, gathered and set aside out of the rain, can be welcome additions to the winter diet. Sheep love apples, even prefer the spoiled ones, and a few apples a day will add needed vitamins.

Molasses is another treat for sheep, and a good source of minerals. As its sugar enters the bloodstream quickly, it is of value to ewes, late in pregnancy, to prevent toxemia.

The pomace from apple cider making is good feed for sheep, if you don't spray your apples. Give it in regular small quantities, while it lasts. Fermented pulp is not harmful if fed in small amounts.

Discarded produce from the grocery is another treat. Lettuce, cabbage, broccoli, celery, and various fruits, past their prime for human food, are often available at the local store. Feed sparingly, or regularly in measured quantities.

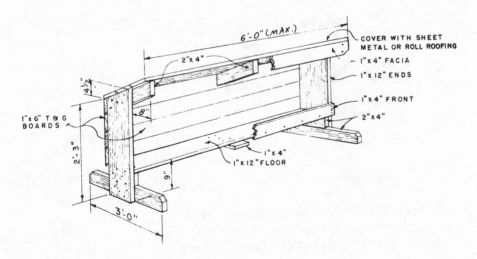

This covered salt box offers protection to the salt from the elements. Plywood can be substituted for the 1" x 6" boards. (Utah State University)

WATER & SALT

Hot summer pasture has very little moisture, so sheep need more water, not just because of the heat, but because their feed contains less water. To cope with the heat, sheep lose more moisture through their skin, and hence they need more water. Providing shade for sheep helps keep down the moisture loss, but they still need clean, fresh water.

Salt is another year-round necessity for good health. When sheep have been deprived of salt for any length of time and then get access to it, they may overindulge and suffer salt-poisoning. (Treatment: access to plenty of fresh water.) Avoid this illness by keeping salt available at all times, preferably with the addition of one part phenothiazine to ten parts salt to assist in the control of internal parasites.

Regular access to salt is said to be useful, along with roughage, to prevent bloat, which is the most serious of the digestive upsets.

BLOAT

Bloat is an excessive accumulation of gas and/or foamy material in the rumen. Severe cases can be fatal in as little as two hours if not treated.

Too much of almost any feed can cause bloat, but overconsumption of wet clover, grain, orchard fruit or lush pasture is the most common cause.

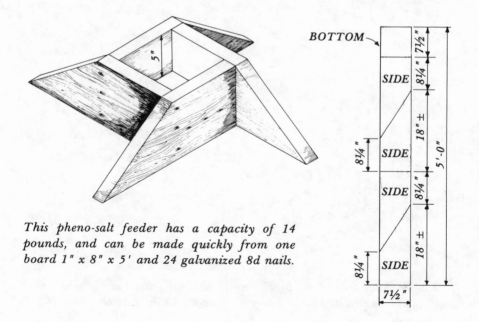

This pheno-salt feeder has a capacity of 14 pounds, and can be made quickly from one board 1" x 8" x 5' and 24 galvanized 8d nails.

Legume pastures, such as very leafy alfalfa and clover, are even more dangerous than grass.

When changed from sparse to lush pasture, sheep may gorge themselves *unless* given a feeding of dry hay prior to turning out on the new pasture. Sheep seldom bloat when they are getting dry hay with their pasture. The coarse feed is thought to stimulate the belching mechanism, as well as keep-

Sheep and Lambing Shed for 30-36 Ewes.

This sheep barn has a grain room, space for use of the hinged lambing pens, and a lamb creep area with a removable creep entry panel so hay can be stored there when area isn't in use for lambs. The feed storage room is safe and secure from the sheep. Store grain in rodent-proof containers. The building does not have to be heavily insulated, but should be tight enough to prevent drafts. If electricity is available, circuits should be provided for heat lamps when needed for newborn lambs.

The building is 20 feet deep, 32 feet long and provides space for about 30 ewes. The length of the building can be increased in multiples of 8 feet. It needs no foundation and creosoted timbers are used. This is USDA Plan #5919, available in two pages of blueprints at approximately $1 per page. (USDA plan)

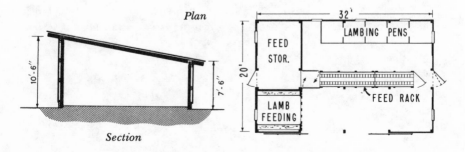

Plan

Section

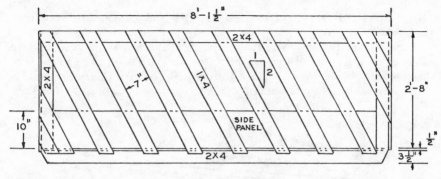

Side view

This is a good hay feeder for the sheep barn shown. These slanted slats discourage the sheep from stepping back with a mouthful of hay and dropping some on the ground.

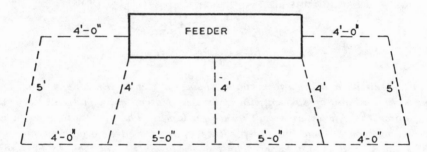

Lambing pens around feeder, up to 8 pens per unit

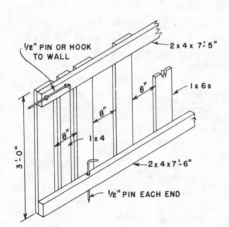

A removable creep fence for lamb feeding area inside barn. (USDA plan)

72

ing the green feed from making a compact mass. Some sheep seem more prone to bloat than others, possibly due to a faulty belching mechanism.

SYMPTOMS

Enlargement of the rumen on the left flank is a sign of bloat. The animal breathes hard, grinds its teeth because of abdominal gas pain, sometimes salivates profusely, and stops eating. When the sheep falls to the ground, death usually follows, probably from suffocation.

TREATMENT

Simple treatments are more likely to succeed if bloat is noticed before it becomes too severe. There are preparations on the market for treating bloat, which are good to have on hand in case they are needed. *(Therabloat and Bloat Guard are good.)*

If bloat is not so severe as to have caused a breathing disturbance, you can prevent further gas formation by giving two tablespoons of baking soda in a cup of warm water, using a dose syringe. Do not raise the sheep's nose above its eye level when doing this, or the mixture may go into the lungs.

Repeat the dosage in thirty minutes if necessary.

You can place the sheep in a sitting position, and massage the abdomen to encourage belching. Another old remedy was to tie a stick between the jaws. To be effective, the sheep should be forcibly exercised when doing this, occasionally raising its front feet off the ground to encourage the release of gas. This is of little value if the bloat is of the foamy type. Foam, which cannot be belched up, can be helped by another one-half cup of vegetable oil (peanut oil or corn oil are recommended) given by mouth if ewe is still able to breathe and swallow normally.

A $\frac{3}{8}$-inch or $\frac{1}{2}$-inch rubber tubing (small siphon tube) down the throat into the stomach can release gas, unless there is too much foam. If you use the finger test to be sure it is not in the lungs,* you can break up the foam somewhat by pouring $\frac{1}{3}$ to $\frac{1}{2}$ cup of vegetable oil into the tube with a funnel.

In an emergency, the rumen can be punctured, preferably by a vet, with proper equipment to relieve both foam and gas, and to treat to prevent infection.

---

*Finger test to determine if a tube is in the stomach where it belongs or in the lungs: Wet a finger and hold it in front of the protruding end of tube. If you feel cool air, like breathing, you are in the lungs. Pull out the tube and try again.

After pouring oil down the tube, withdraw it fairly quickly, to avoid dribbling any oil at the entrance of the lungs.

CHAPTER FIVE

THE RAM

The ram is half of your flock, so inspect him well before buying. The following points should be kept in mind when looking at a ram for possible purchase:

1. Good size, deep wide body, heavy muscular neck.
2. Well-developed sex organs, two normal testicles.
3. No scrotal mange, no hernia. Turn ram up to detect.
4. Good feet. Bad feet can render ram useless.
5. Good eyes. Watch for pink eye or any sign of eye damage.
6. Good teeth, well-aligned with upper jaw.
7. Head not too large, to cause hereditary problems in lambing.
8. Full hindquarters, so his lambs will be good meat animals.
9. Good fleece, of type to suit your wool market.
10. General health of the rest of the flock.

Ordinarily the best investment is a well-grown two-year-old, a twin or a triplet. Being a twin will in no way affect the twinning of the ewes he breeds. This is controlled by the number of eggs the ewe drops to be fertilized, which is influenced by genetics and encouraged by flushing. However, the lambs they have will inherit both the ewe's *and* the ram's twinning capabilities, and this will show up in the following generation. The ram greatly influences body conformation and fleece type.

Buy your ram long enough before breeding season so he becomes acclimated to his new home. Keep him separate on good feed and pasture until breeding time. Check his hooves and trim them if they need it; worm him. Rams should be sheared rather late in the spring so they will have a short fleece in the hot weather, as heat lowers their fertility. Excess weight also results in a lowering of potency and efficiency, so keep him in good condition, but not fat.

If he is a lamb, use him sparingly for breeding. One good ram can handle twenty-five to thirty ewes. On a small farm where the ram gets good feed, you can expect about six years of use from him. On open range, where there may be over-use with more ewes per ram, and little supplemental feed, rams get run down during the breeding season from eating so little and chasing the girls. They then succumb to other diseases because of their low resistance.

During the breeding season, feed the ram about one pound of grain per day, so that if he is too intent on the ewes to graze properly, he is still well nourished.

After separating him from the bred ewes, a maintenance ration of one-half pound of grain per day, plus hay as necessary during the winter, should carry the ram through until good pasture is available again.

Provide a cool shady place for him in the heat of summer. Semen quality is affected at 80 degrees, and seriously damaged at 90 degrees. Several hours at that temperature may leave him infertile for weeks, and cancel your plans for early lambing. If your climate is very hot in the summer, shear his scrotum just before the hot weather, and run him with the ewes in the evening, at night, and in the early morning, but keep him penned in a cool place during the heat of the afternoon.

August generally is the beginning of breeding season for early (January) lambing. You can wait until later to turn the ram in with the ewes, if you want to start lambing later in the spring. Gestation is five months (148 to 152 days).

Ewes are in heat about twenty-eight hours, with about sixteen to seventeen days between cycles, so fifty-one to sixty days with the ram should mate all the ewes, even the yearlings who are sometimes late in coming in heat.

A newly sheared Suffolk ram, owned by E. William Hess, Barboursville Farms.

The sense of smell greatly determines a ram's awareness of estrus in the ewes. A study of sex drive in rams, done at University College in Aberystwyth, Wales, found that some breeds of rams have keener olfactory development than others, and are able to detect early estrus in ewes that goes unnoticed by other breeds. Those with the "best noses" for it were singled out as Kerry Hill, Hampshire, and Suffolk rams, in that order.

RAM MARKING HARNESS

To keep track of the ewes who are bred, and when, a "marking harness" is available in many sheep supply catalogs for use on your ram. It has a holder on his chest for a marking crayon. Each ewe is marked with the color of the crayon the day he breeds her. Inspect the ewes each day. Keep track of the dates so you will know when to expect each one to lamb. Use one color for the first sixteen days he is with the ewes, then change color for the next sixteen days, and again for the next. If many ewes are being re-marked, it means they were not bred the previous times he tried to breed them, as they are still coming into heat, so you may have a sterile ram.

If the weather was extremely hot just before or after you turned him in, you can blame it on the heat. But to be safe, you should get another ram in with them, in case your ram's infertility is not just a temporary one caused by hot weather.

PAINTED BRISKET

Instead of a purchased harness, you can daub marking paint on the ram's brisket (lower chest). Mix it into a paste with lubricating oil, or even vegetable shortening, using only paints that will wash out of the fleece. Suggested colors to use: Venetian red, lamp black, yellow ochre.

RAISING YOUR OWN RAM

If you are raising market lambs for meat, you might try a system called "recurrent selection of ram lambs," which consists of keeping the fastest-gaining ram lambs sired by the fastest-gaining ram lambs. No, this is not a misprint. Recurrent selection of ram lambs is a way of improving the potential for fast growth in your lamb crop. It involves changing rams fairly frequently, and leaves you the problem of disposing of a two- or three-year-old ram. If he is a good one, you can probably sell him as a ram; if not, see chapter on Muttonburger.

When you are raising a lamb for a breeding ram, do not pet him or handle him unnecessarily. Do not let children play with him, even when he is small. He will be more prone to butting and becoming dangerous if he is familiar with you than if he is shy, even a little afraid of humans.

One advantage of raising your own ram is that you see what he looks like when he is at what would be market age if he were sold for meat. The older a ram gets, the less you can tell about how he looked as a lamb, or how his offspring will look when they are market age.

THE BUTTING RAM

I wanted to call this "the battering ram," but in actual cases, it is not very funny.

Keep children away from a ram. He may hurt them badly, and they can make him playful and dangerous. Never pet him on top of his head; this encourages him to butt.

Leading a ram with one hand under his chin will keep him from getting his head down into butting position. A ram butts from the top of his head, not from the forehead. His head is held so low that as he charges you, he does not see forward well enough to swerve suddenly. A quick step to the right or the left will let you avoid the collision.

If you have a ram who already butts at you, try the water cure: a half pail of water in his face when he comes at you. After that, a water pistol or dose syringe of water in his face usually suffices to reinforce the training.

Author is shown being chased by her least favorite ram, and, seconds later, being rescued by her husband. (Washington State Department of Commerce and Economic Development)

If you have a dangerous ram who is very valuable, he can be hooded so that he can only see a bit downward and backward. He must then be kept apart from other rams, as he is quite helpless.

Strange rams will fight when put together. Well-acquainted ones will too, if they've been separated for a while. They back up and charge at each other with their heads down. If you have two strong rams who are both very determined, they will keep at it until their heads are bleeding, and one finally staggers to his knees and has a hard time getting up. Once they have determined which one is boss, they may playfully butt, but will fight no big battles unless they are separated for a time.

To prevent fighting, and the possibility of one ram's being badly injured, you can put them together in a small pen for a few days at first. In a confined area, they can't back up far enough to do any damage.

If no pen is available, they can be "hoppled," "yoked," or "clogged" — all old European practices.

Hoppling a ram (the modern term would be *hobbling)* was an old system of fastening the ends of a broad leather strap to a fore and a hind leg, just above the pastern joints, leaving the legs at about the natural distance apart. It discourages rams from butting each other, or people, because they are unable to charge from any distance, and they cannot hit very hard if they can't get a run. They may stand close and push each other around, but will do nothing drastic. Hoppling also keeps them from jumping a fence, which rams will sometimes do if ewes are in the adjoining pasture.

Yoking is fastening two rams together, two or three feet apart, by bows or straps around their necks, fastened to a light timber like a 2-inch x 3-inch piece of lumber.

Either of these methods will need watching, so that the rams do not get entangled.

In *clogging,* you fasten a piece of wood to one fore leg by a leather strap. This will slow the ram down and discourage fence jumping or fighting. Close watching is not necessary.

CHAPTER SIX

FLUSHING AND BREEDING

Sheep operations revolve around the growth cycle of the pasture, so the desired month of lambing will depend on pasture growth. It takes the rumen of newborn lambs about six weeks to develop, or a little less if creep feeding starts as early as ten days old. So plan for the lambs to be about five–six weeks old about the time of the first good early growth of pasture. Plan to breed about five months before you want lambs.

Here are the advantages of both early and late lambing:

EARLY LAMBING

1. There are fewer parasites on the early grass pasture.
2. Ewe lambs born early are more apt to breed as lambs.
3. You can sell early lambs by Easter, if creep fed, and get a better price for the early meat lambs.
4. You can have all lambs born by the time of the best of the spring grass.
5. There are fewer problems with flies at docking and castrating.

LATE LAMBING

1. It's easy to shear ewes before lambing.
2. There isn't as much danger in lambing if winters are severe.
3. Mild weather means fewer chilled lambs.
4. Ewes can lamb out on the pasture.
5. Less grain is required for lambs, since you have lots of pasture.

GET READY TO START BREEDING

Worm your ewes, and trim away any wool tags from around the tail. Trim their feet, for they will be carrying extra weight during pregnancy and it is important that their feet be in good condition. Worm the ram too, and check all for ticks. If you eliminate ticks before lambing, none will get on the lambs, and you will not have to treat for ticks again.

At seventeen days before you want to start breeding, put your ram in a pasture adjacent to the ewes, with a good fence between them. Australian research has determined the sound and smell of the ram will bring ewes into heat earlier.

FLUSHING

Flushing is proven most effective when done for seventeen days prior to turning the ram in with the ewes, and no advantage is shown by starting it earlier. Flushing, or supplementing the usual summer diet with grain (and sometimes a better pasture, too), is done just before breeding, for it not only gets the ewes in better physical condition for breeding, but results in bringing all the ewes into heat at about the same time, which prevents long, strung-out lambing seasons.

It is also a factor in twinning, possibly because with this better nourishment the ewes are more likely to drop two ova. The USDA estimates an 18 percent to 25 percent increase in the number of lambs by flushing, and many farmers think it is even more.

You can start with $\frac{1}{4}$ pound of grain a day per ewe, and work up to $\frac{1}{2}$ or $\frac{3}{4}$ pound each in the first week. Continue at that quantity for the seventeen days of flushing. When you turn in the ram, do not abruptly discontinue the grain, but taper it off gradually, taking about two weeks.

Suffolk x Dorset ram, in "courtship" pose. (Photo by ram's owner, Jim Gloe)

The ewes will probably come into heat once during that seventeen days of flushing, but it is best not to have the ram with them then, for in the second heat they drop a greater number of eggs, and are more likely to twin if bred during this second cycle.

Ewes should not be pastured on red clover, as it contains estrogen and lowers lambing percentages. White clover may have somewhat the same effect.

EWE LAMBS

The exception to the flushing would be the ewe lambs, if you decide to breed them. They will not have reached full size by lambing time, so you would not want them to be bred too early in the breeding season. Don't breed them until the following month. Breeding season is shorter for ewe lambs than for mature ewes, and usually starts in September or October, instead of August. Some breeds are slower maturing, like Rambouillet, and some much faster, like Finnsheep and the new Polypay.

Ewe lambs should have attained a weight of 85 to 100 pounds by breeding time, as their growth will be held back a little as compared to unbred lambs. If not well fed, their reproductive lifetime may be shortened, and unless they get a mineral supplement (like TM salt), they will have teeth problems at an earlier age.

If replacement ewes are chosen for their ability to breed as lambs, the flock will improve in the capacity for ewe lamb breeding, which can be a sales factor to stress when selling breeding stock.

CHOOSE TWINS

Choose your potential replacement ewes from among your earlier born twin ewes. Turn these twin ewe lambs in with a ram wearing a marking harness, or paint-marked brisket (see Chapter 5). The ones that are marked, and presumably bred, can be kept for your own flock. Sell the rest.

> *Ewes yearly by twinning*
> *Rich masters do make*
> *The lambs from such twinners**
> *For breeders do take.*

Youatt, 1837

Ewes who breed as lambs are thought to be the most promising, as they show early maturing which is the key to prolific lambing. A ewe lamb that has twins the first time is more valuable than one who lambs with a single, even though ewes with a future history of twinning may only have a single that first time. Still, they pass on both the inherited ability to breed early and to have twins, and they will produce more lambs during their lifetime.

---

*And, I might add . . . who are bred as lambs.

PUREBREDS AND CROSSBREEDS

When both parents are purebreds of the same breed, the lamb is also a purebred.

When each parent is of a distinct different breed, the lamb is a crossbreed. In crossbreeding you get a lamb that can potentially, but not necessarily, have the good points of both of the parents, and is usually faster growing. The value of crossbreeding can be determined in practice by comparing the lamb with the two parent breeds, considering particularly the factors that are of importance in your situation: body conformation or wool or prolificacy or rate of growth, or size.

GRADING UP

The use of a good purebred ram on a flock of very ordinary ewes, and keeping the best of the offspring, is called "grading up." If done for several years, keeping the best of the resulting ewe lambs, and disposing of the original ewes, you have probably improved the quality of your flock. The actual improvement will depend partly on the ram chosen, and partly on how carefully you select the ewe lambs that you keep for replacements.

BACKCROSSING

With an unusually good ram, you may want to go one step further, breeding these ewe lambs back to the same ram, which is called "backcrossing," a form of inbreeding. The lambs resulting from this mating should not be bred back to the same ram.

Inbreeding or close breeding, once referred to as breeding in-and-in, is not without risks, but you can always cull out any lambs that show undesirable traits, such as being undersized, the most common fault of inbreeding.

LINEBREEDING

Linebreeding is a form of inbreeding in which sheep are mated in such a way that their lambs will remain closely related to one highly desirable ancestor. The difference from common inbreeding is that the mated animals should both be related to the one unusual ancestor, but *unrelated* to each other. The intent is to maintain a close relationship to that one particular outstanding animal, and propagate its exceptional characteristics, not allowing them to be halved in each following generation.

EXPERIMENTS WITH INBREEDING

The McMaster Field Station in Australia did tests in 1965 to determine the cause of the most common weakness of inbreeding—a reduction in growth rate and loss of vigor. The tests placed part of the blame on a decreased activity of the growth-controlling hormone in inbred lambs.

The scientists were able to reverse this by giving the lambs injections of pituitary gland extract during the first ten weeks after birth. The inbred lambs actually grew faster and more vigorously than the control group of non-inbred lambs. Continuation of the injections after the first ten weeks produced no further benefits, but the differences in body size and weight continued long after the treatment.

CULLING

By keeping the best of your ewe lambs, and gradually using them to re-place older ewes, you should realize more profit.

To know which to cull, you need to keep good records, and this necessi-tates ear tags. You're more inclined to keep clear records with tags than without, and you can also be more efficient about them. Record: Fleece weight each year, lambing record, rejected lambs, milking ability, lamb growth, any foot troubles, and illnesses and treatment.

With a good history of the animal, you know better what to anticipate. At culling time, review the records, as well as inspecting teeth, udders, and feet. Cull out ewes with defective udders, broken mouth (teeth missing), limpers who do not respond to regular trimming and foot baths, or those with insufficient milk, whose lambs grow slowly.

TWO LAMB CROPS PER YEAR

The profitable possibility of attaining two lamb crops a year is now much closer because of work done by the Animal Science Department at Utah State University. Scientists there devised and tested a method to overcome the common problem of uterine debris that prevents ewes from breeding back early enough to have two lamb crops in twelve months.

Dr. Warren C. Foote explained that the infusing of the uterus with 200 ml saturated sucrose solution* via the cervix, within four days after lambing, obtained beneficial response. He said that this definitely proved to be effective in preparing the ewes to breed. Sterile solution and a sterile procedure would be very necessary to avoid serious complications.

The breeds most noted for out-of-season breeding are the most likely candidates for the practical application of this method. The new Polypay breed was developed specifically for the feature of twice-a-year lambing, so they could be at the top of the list, needing only a little more help to be reliable double producers. Dorset and Finnsheep are good for out-of-season breeding, and those with a moderate capacity for it would be Rambouillet, Debouillet, Romney, Targhee, Tunis, Panama, Merino, Hampshire, and Lincoln.

The breed involved in the Utah tests was a Targhee type.

---

*Concentrated sucrose solution is a form of sugar and water, obtained by stirring sugar into boiling water, adding as much sugar as will dissolve in it. When cooled, the liquid decanted off the top is concentrated sucrose solution.

CHAPTER SEVEN

PRE-LAMBING AND LAMBING

FEEDS

Do not overfeed ewes during the early months of pregnancy. A program of increased feeding must be maintained during late gestation to avoid pregnancy disease and other problems. Overfeeding early in pregnancy can cause ewes to gain weight and they will have difficulty in lambing.

Have adequate feeder space so that all the ewes will have access to the feed at one time; otherwise, timid or older ewes will get crowded out. If possible, they should have free choice of a mineral supplement in loose form separate from the salt. Also, a low-level phenothiazine-salt mix, ratio of one part pheno to ten parts salt, should be provided.

FEEDING IN LAST FOUR OR FIVE WEEKS
BEFORE LAMBING

By the fourth month of pregnancy, ewes need about four times as much water as they did before pregnancy. And, since 70 percent of the growth of unborn lambs takes place in this last five-six weeks, their feed must be adequate to support that growth. Since by that time the lamb embryos are so large that they displace some of the stomach space, the ewes are not able to handle as much roughage, or large quantities of any low-nutritive feed. Increase their grain rather than their hay. A good grain mix would be $\frac{1}{3}$ whole oats, $\frac{1}{3}$ shelled corn, and $\frac{1}{3}$ wheat (for the selenium content). Grain and hay should be given on a regular schedule, to avoid the risk of triggering pregnancy disease by erratic eating.

Poor feeding in last four weeks (last five–six weeks for twinning ewes) leads to:

1. Low birth weight of lambs.
2. Low fat reserve in newborn lamb, resulting in more deaths from chilling and exposure.
3. Low wool production from those lambs as adults.

4. Increased chances of pregnancy toxemia.
5. Shortened gestation period, some born slightly premature.
6. Ewes slower to come into milk, and less milk.
7. Production of "tender" layer in ewe's fleece. This is a weakness that causes the fibers to break in two with the slightest pull, and decreases the wool value.

At this time, watch for droopy ewes, ones going off their feed or standing around in a daze. See Chapter 12 for symptoms and treatment of pregnancy toxemia. Both exercise and sunlight are valuable to a ewe that is carrying a lamb, and lack of exercise is one factor in pregnancy disease. If necessary, spread hay for them in various places on clean parts of pasture, once a day, to get them out and walking around.

SHEARING BEFORE LAMBING

If weather is mild and you do your own shearing so you can be gentle with them, ewes can be sheared up to three or four weeks before lambing. See Chapter 17 for shearing. There are some advantages in having ewes sheared before lambing:

1. No dirty, germ-laden wool tags for lamb to suck,
2. Clean udder makes it easier for lamb to find teats,
3. Fewer germs in contact with lamb as it emerges at birth,
4. Easier to assist at lambing if necessary,
5. Easier to spot an impending prolapse, in time to save ewe (see Prolapse, Chapter 12),
6. Easier to predict lambing time by ewe's appearance,
7. Ewe less apt to lie on her lamb in pen,
8. Shorn ewe requires less space in barn, at feeding racks, and in lambing pen,
9. Shorn ewe not as apt to sweat in lamb pen and contract pneumonia,
10. Shorn ewe will seek shelter for herself (and lamb) in bad weather.

CROTCHING BEFORE LAMBING

Actually, the first five advantages are gained also by *crotching* (sometimes called *crutching)*, which is trimming wool from the crotch and udder, and a few inches forward of the udder on the stomach. Only about four–five ounces of wool are removed, and this can be washed for use if you spin, or sold with the fleece.

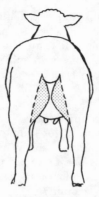

*Darkened area is area
for crotching.*

FACING
(called "wigging" in some countries)

Another practice of value before lambing, or before ewe is turned out of the lambing pen, is "facing," (trimming the wool off the ewe's face). It has two purposes. In heavy-wooled sheep, it enables the ewe to locate and watch her lamb more easily. It also avoids the tendency to "wool blindness" which can inhibit feed consumption. With lighter-wooled sheep, it still makes the ewe more prone to seek shelter with her lamb in wind and rain, even if she has not been sheared.

LAMBING SYMPTOMS

As the time approaches for actual lambing, the ewe will appear restless and sunken in front of the hip bones. This is much more noticeable if she has been sheared. She will often lie down apart from the rest of the sheep, sometimes pawing the ground before lying down. Too much lying around without any observable cud-chewing may be an early sign of the droopiness of toxemia (see Chapter 12.) She usually will have made a bag by now, but some make more than others. She may refuse a grain feeding just before lambing, but will gobble up an apple cut in pieces. Our ewes who are huge with twins or triplets start grunting several days before lambing, as they lie down or get up. Often the vulva will be a little pinker than before—but should not be protruding and red. This would be the beginning of prolapse (see Chapter 12).

LAMBING PEN

Have a 4 x 5 foot (or 5 x 5 for large breed sheep) lambing pen or "claiming pen" ready for the newborn lamb and mother, with clean bedding, a small hay feeder, and a container of water that cannot be spilled.

Ewes lambing for the first time, especially yearlings, should be penned with their lambs until they become accustomed to nursing them, as they do not have the precedent of previous lambings, or the fully developed maternal instincts.

If a young ewe does not at once have enough milk for the lamb, supplement it with a couple of two–ounce bottle feedings for the first two days, with milk taken from another ewe, or with newborn milk formula (see Chapter 10).

Her milk should increase if she is well fed. If it still is not sufficient for the lamb, supplement it with a couple of four-ounce feedings of lamb milk replacer during the first week, then increase to about eight-ounce feedings at two weeks old. Poorly fed old ewes also may have scant milk supply. If the lamb cries a lot, that is one indication that it is hungry.

The pen is primarily for use after the lamb is born. Ewes prefer a larger area for lambing, where they can walk around freely before labor. The larger pen (5 x 5 feet) is best if you want to have the ewe confined where facilities are better for helping in a difficult birth. If weather is bad, you may want to have the ewe in the pen as labor starts, where there is good light to watch her progress.

If she does lamb outside, it is not difficult to get her to the claiming pen nearby. Carry the lamb slowly so that she can see it and follow. Sometimes she runs back to where she had it, and you have to go back and start all over. If the lamb calls out to her along the way, she follows readily.

The pen allows the ewe and her lamb to become well acquainted without distraction, keeps the lamb from getting separated from its mother, and protects it from being trampled by other sheep or becoming wet and

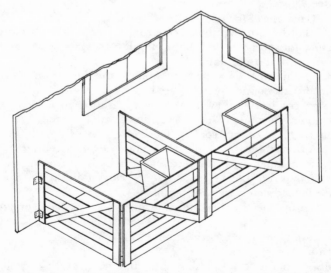

Folding lambing pens, in place in the barn. Ewe hay racks are shown in place.

chilled. Ordinarily, the two are penned for three days so that they can be easily observed and treated, should complications arise.

LAMBING

Keep your fingernails trimmed close, in case you have to assist in delivery, and have the following pre-lambing and lambing supplies on hand. Many are obtained by mail. Order them ahead of time so you will have them when needed. Here is a list, not necessarily in the exact order of importance or in the sequence they may be needed:

1. Roll of paper towels in lambing pen.
2. Iodine (or Lugol's solution) for treating umbilical cord.
3. Hand sheep shears for crotching. Can also be used for annual shearing.
4. Antiseptic and lubricating ointment for your hands, if you have to assist in delivery. *Cooper's Diary Ointment* or *Septi-Lube* are good.
5. Antibiotic uterine boluses in case of retained afterbirth.
6. Penicillin, sterile syringe and #18 needle.
7. Lambing "snare" to help pull out lamb in difficult delivery.
8. Heavy twine for lambing loops (snares). Dip in antiseptic before using.
9. Livestock molasses. (Grocery store kind is too expensive; get this at feed store.)
10. Propylene glycol for treatment of pregnancy toxemia.
11. Baby bottle with slightly enlarged nipple hole. This is more satisfactory for *newborn* than lamb nipples; they are too large for the newborn, but useful later.
12. Heat lamp with ceramic-base holder and heavy duty extension cord if pens not adequately wired.
13. Frozen colostrum, thaw at room temperature if needed, *or* newborn lamb formula (Chapter 10), make up if needed.
14. Mineral oil in case of constipated lamb.
15. Pepto Bismol for simple diarrhea caused by overfeeding.
16. Elastrator pliers with rubber rings, for both castration and docking (removal of tails), see Chapter 9. Dock tails at two–three days old, depending on vigor of lambs.
17. Thibenzole sheep boluses for internal parasites. Safe for pregnant ewes.
18. Calcium gluconate for treatment of milk fever.
19. Lamb ear tags.
20. Sulfa preparation for lamb scours (diarrhea).
21. Small scissors.
22. Bucket of warm water.

START OF ACTUAL LAMBING

Labor starts when the ewe lies down with her nose pointed up, then strains and grunts. Give her plenty of time to lamb by herself before trying to assist, unless she gets the lamb half way out and is making little progress. You can pull on the lamb, timing your pulls with her straining.

Most vets say allow ½ to 1 hour after the water bag comes out, or 1½ to 2 hours of labor without the lamb appearing, but you'll have to judge from her appearance as to whether the ewe is becoming so tired that you should check for the lamb position and help her.

In the great majority of cases, she will give birth normally and easily. If you have any lambing problems, see Chapter 8 for instructions on how to deliver lambs.

At birth. If you are there when the lamb is born, wipe the mucus off lamb's nose, then place it at the ewe's head quickly, so she can identify it as her own and clean it off. (Now would be the time to "graft" on an orphan or quadruplet that needs a foster mother. See instructions in Chapter 10.)

If the navel cord is over 2 inches long, knot it carefully about 1 inch to 1½ inches from the body and snip it off below the knot. Then apply 7 percent iodine diluted with a little alcohol, or substitute Lugol's solution if you raise horses and have it for them, or are in a known tetanus area. Have the iodine in a small wide-mouthed container. Hold the lamb so that the navel cord hangs into the container. Press the container against the lamb's belly, then turn lamb up so that the entire cord and the area surrounding it are covered. Iodine should be applied as soon as possible after birth, as many germs can enter via the navel. The iodine penetrates the cord, disinfecting it, and assists in drying it. Some ewes try to nibble too much on the navel if the cord is left too long, and can injure the lamb.

This year we had an excited ewe who chewed the tail off her newly born lamb, nibbling it as if it were the umbilical cord. In over twenty years, this had never happened. We put a band on the tail, above where she had chewed it, and dunked it in iodine, then went to get warm molasses water for her to drink. When we got back, she had just had a second lamb, and we couldn't believe it. She chomped the tail off that one, too! Nice that she was in the lambing pen with good light, so we could see what had happened, and could take care of those poor tails. She licked off the lambs, and is a wonderful mother. If this had happened with more than one ewe, we would have suspected a nutritional deficiency, for that is reported to be one sign of it.

If the ewe is too exhausted by a difficult labor to dry off the lamb, help her with paper towels, so that the lamb does not get cold from being wet too long. But do not remove the lamb from her sight, as this can disrupt the mothering-ownership pattern. Allow her to lick off the amniotic fluid, but even if she is not able to, put the lamb near her nose to encourage her to identify with it. Overuse of heat lamp to dry the lamb, if unnecessary, will predispose it to pneumonia later.

The author's Lambie Pie and one-day-old lamb.

If the lamb should be born dead, you can rub a young orphan lamb all over with the birth fluid and give it to the ewe to mother. (See Chapter 10.)

Unplug the ewe. Strip the teats of the ewe to unplug her, as the lamb may not suck strongly enough to remove the little waxy plug.

Check eyelids. Check the lamb's eyelids to see if they appear to be turned in so that the eyelashes would irritate the eye. This can cause trouble if it is not corrected, and the sooner it is noticed the easier the remedy. (See Entropion, Chapter 11.)

NURSING

When the ewe stands up, she will nudge the lamb toward her udder with her nose, if it is strong enough to get on its feet. The lamb is born with the instinct to look for her teats, and also is drawn by smell from the waxy secretion of the mammary pouch gland in her groin. Some people swab the ewe's udder and teats with a mild chlorine solution before the lamb nurses.

Let the lamb nurse by itself if it will, but do not let more than a half-hour pass without it nursing, as the colostrum (ewe's first milk after lambing) provides not only warmth and energy, but also antibodies to the specific germs of its environment. These antibodies are absorbed by the mammary gland from the ewe's blood and passed into the colostrum so that they protect the newborn lamb until it starts to manufacture its own antibodies. The small intestine of the newborn possesses the very temporary ability to absorb these antibodies from the colostrum. Colostrum is also high in vitamins and protein, and is a mild laxative to pass the fetal dung (the black tarry substance that comes out shortly after the lamb nurses).

The longer a lamb has to survive without colostrum, the less its chance of survival if it develops problems. A weak lamb or one of light birth weight can be lost because of a delay in nursing.

This is not the same as loss due to starvation, or from receiving no milk at all, as a strong lamb can sometimes survive for a day or more without ever getting any milk, but getting weaker all the time. Many lamb deaths that are attributed to disease are actually starvation.

We don't usually wait around for the lamb to nurse, but just roll the ewe on her side and press the lamb's nose against her teat. If it does not readily suck by itself when it feels the warm udder, we push the teat into its mouth, from the side. It usually cooperates, getting the urge when it feels the warmth in its mouth. After this first feeding, we have some assurance that it will have the strength to look for the next one, but we keep watch to make sure that it does nurse from time to time.

MOLASSES AND FEED FOR MAMA

After she has given birth, offer the ewe a large bucket of warm water (not hot) containing half a cup of stock molasses. It is important to have it warmed, as cold water can cause her milk to hold up. Offer good hay, but no grain the first day, as it could promote more milk than a tiny lamb could use. If she has twins or triplets, however, and seems short of milk, grain feeding could start that first day.

If she has too much milk, udder is too full and teats enlarged from it, milk out a bit of this colostrum and freeze it in small containers for emergency use. It will keep a year if solidly frozen and well wrapped.

When saving and freezing colostrum, it is even better to have a combination of colostrum milked from several ewes, for they do not all produce colostrum with a broad enough spectrum of disease and infection-fighting antibodies.

Note: For another way to harvest antibodies and freeze them, see Blood Serum, end of Chapter 10.

TWINS

Twins require vigilance to assure that both lambs are claimed by the ewe, and that each is getting its share of colostrum. If the ewe does not have plenty of milk for them, increase grain gradually. Unless she shows some reluctance about the molasses, continue offering it in lukewarm water during the time she is penned.

If twins cry a lot, they are probably not getting enough milk. Notice if only one of them is crying, or both, and assist the hungry one by holding it to its mother. If she is short on milk for both, give a supplemental bottle. When a ewe does not have enough milk for multiple lambs, you can still leave them all nursing her, but supplement one of them or all of them with a couple of bottle feedings a day. Give two–ounce feedings the first couple of days, and increase to four–five ounces by the third and fourth day,

Targee and triplets. (Mt. Haggin Livestock, Inc., Anaconda, MT)

gradually increasing as they grow, if her milk is still not adequate. See Chapter 10 for the newborn lamb milk formula to feed for the first two days. Then gradually change to lamb milk replacer (not calf milk replacer).

LAMB DROPPINGS

One advantage of penning lambs with their mothers is that you can keep an eye on how well they are eating, and on how well it is coming out the other end — the condition of the droppings is important.

First to come out is the fetal dung, a gob of black tarry matter, coming out in the first few hours after the lamb is born. The next droppings are bright yellow, but the same consistency. They remain yellow for at least a week, then gradually get darker until they are a normal brown small bunch of pellets sticking together in clumps. Later, they are little brown marbles.

If they become loose and runny, this is called scours. See Chapter 11 for treatment.

EAR TAGS

If you have more than two or three ewes, which should produce two to six lambs, you can identify them best by ear tags. This makes it possible to keep records of lamb parentage, date of birth, and growth, and easier to decide what to keep for your flock and what to sell. With identification tags on your ewes also, you can be certain which lambs are hers, even after they are weaned.

Livestock supply catalogs sell a variety of tags. Some are metal with almost any combination of numbers and letters (your name if you wish), and some are plastic in a variety of colors also with numbers and letters of your choice. The different colors can be used to identify by sex, or whether twins or singles, or the month born, etc. Some are a self-clinching type, while others need a hole punched for the tag. These should be applied while the lamb is still penned with its mother.

CHAPTER EIGHT

ABNORMAL LAMBING POSITIONS AND HOW TO HELP

Usually the ewe will give birth unassisted, but you should be prepared for the abnormal delivery. During lambing season keep your fingernails cut short, in case emergency requires you to reach in to pull a lamb. You also will need the following supplies readily available if it is necessary to help the ewe when lambing.

SUPPLIES TO HAVE ON HAND

1. Good light in the delivery area.
2. A lambing snare, or several pieces of strong cord, with a noose on the end of each one.
3. Antiseptic lubricant or mineral oil.
4. Bucket of clean soapy water to wash your hands and arms, and external parts of the ewe.
5. Penicillin to give after assisting, if in doubt that sterile procedure was used.
6. Roll of paper towels.
7. Iodine in small wide mouth bottle.
8. Antibiotic uterine boluses.

HELPING THE EWE

It is always a quandary to know when to help and when not to help. As a general rule, you can allow a half-hour to an hour after the water bag breaks, or $1\frac{1}{2}$ to 2 hours of labor, before you try to determine the position of the lamb. You want to give her time to expel it herself if she can, but not wait until she has stopped trying.

The size of the pelvic opening is usually large enough for the lamb's body to come out if it is in the normal position, with the front legs and the head coming first. If it is not in this position, delivery is seldom possible without some repositioning of the lamb, or assistance.

After you have washed your hands and arms, and washed off the ewe, you can lubricate one hand and slip it in gently, to try to find out the position of the lamb.

Identify the lamb's legs, and position. First, make sure that the legs you feel belong to the same lamb. In twin births, frequently one or both of the lambs come backwards, and it's easy to get their legs mixed up.

The front legs, above the knees, have a muscular development. The hind legs have a prominent tendon. The front knee bends the same way as the foot (pastern) joint, with the knuckle pointing forward. The back knee joint bends the opposite way from the back foot, and has a sharper knuckle, pointing backwards. If you have a small lamb, catch it and feel the difference between its front and back legs.

When repositioning a lamb to change an abnormal position, avoid breaking the naval cord, as the lamb will attempt to breathe as soon as the cord is broken.

When helping, time your pulling to coordinate with the ewe's labor contractions. If she has tired and has stopped trying, she will usually start again when you start pulling on the lamb.

After difficult lambing. Place the lamb at the ewe's nose. She may be exhausted and otherwise might not clean the lamb, an act which reinforces the mothering instinct. If her nose feels that warm, wet lamb, she will usually try to lick it off. You may have to help her get it dry. Then, put iodine on its naval, and assist it to nurse, even if the ewe cannot stand up yet.

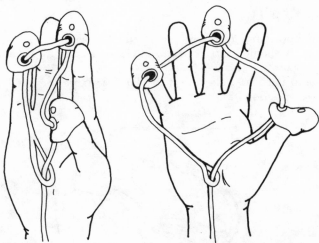

These soft "sensory fingertips" are placed on your own lambing-loop, and are a great help in positioning loop over the lamb's head, for an assisted delivery. They are very popular in New Zealand, where they are made. To order, see Sources, Appendix A.

When to call the vet. If a ewe is obviously in distress, has labored over an hour with no progress, and you cannot get the lamb into proper position for delivery, call the vet. When you are paying for a vet, be sure you learn all you can. They don't ordinarily explain things unless you ask questions, and show an interest.

If a lamb is dead in a ewe, and so large it can't be pulled out, a veterinarian may have to dismember the lamb to remove it.

POSSIBLE LAMB POSITIONS

1. Normal, front feet and head coming out.
2. Large head or shoulders (tight delivery).
3. Front half of lamb out, hips locked.
4. Head and one leg, with one leg turned back.
5. Head, with both legs turned back.
6. Both legs, with head turned back.
7. Hind feet coming first.
8. Breech.
9. Lamb lying crossways.
10. All four legs presented at once.
11. Twins, mixed up, presented at once.
12. Twins, one coming backward, one forward.

1. NORMAL BIRTH

Nose, and both front feet, are presented. The lamb's back is toward the ewe's back. It should start to come out a half-hour to an hour after the ewe has passed the water bag.

She should need no help unless the lamb is large, or has large head or shoulder (see position #2).

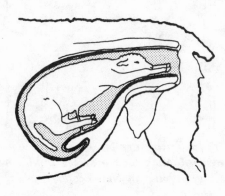

2. LARGE HEAD OR SHOULDERS, TIGHT DELIVERY

Ewe may have difficulty lambing, even with the lamb in normal position, if the lamb is extra large, or the ewe has a small pelvic opening.

Sometimes the shoulders are large, and stopped by the pelvic opening. Use a gentle outward and downward pulling. Pull to the left or the right, so shoulders go through at more of an angle, and more easily.

Occasionally the head is large, or swollen if the ewe has been in labor quite a while. Assist by pushing the skin of the vulva back over the head. When the lamb is half-way out (past the rib cage), the mother usually can expel it by herself, unless she already is exhausted.

When the head is extra large, draw out one leg a little more than the other, while working the ewe's skin back past the top of the lamb's head. Once the head is through, you can extend the other leg completely, and pull out lamb by both legs and neck. If both of the legs are pulled out equally, the thickest part of the legs comes right beside the head, making delivery more difficult (on ewe and you).

Pulling gently from side to side assists birth more than only outward and downward movement as in normal delivery.

Use mineral oil, or antiseptic lubricant with difficult, large lamb. Use loop over lamb's head so that the top of the noose is behind the ears, and the bottom of the snare is in the lamb's mouth. Gentle pulling on the head as well as the legs, is better than pulling on legs only.

3. FRONT HALF OF LAMB OUT, HIPS STUCK

This is only a difficult position for the ewe, who may be weary from labor and need help. While pulling gently on the lamb, swing it a bit from side to side, and if this doesn't make it slip out easily, give it about a quarter turn, while pulling. A large lamb in a small ewe will often need this kind of assistance.

4. HEAD AND ONE LEG COMING OUT

Veterinarians are not in agreement on the procedure to follow when the head and one leg are presented, with one leg still turned back. Some reason that since some ewes can lamb unassisted with the lamb in this position, it should be safe enough to pull gently on the one leg and head, to deliver the lamb, if the ewe is having difficulty. Others says it's risky, and that the folded-back leg may kick at the ewe, causing internal damage.

To change this to a normal birth position, attach a snare-cord to the one leg that is coming out, and also one onto the head. Then push them back enough to enable you to bring the retained leg forward, so you can pull the lamb out in normal position. The cord on the head is important, for the

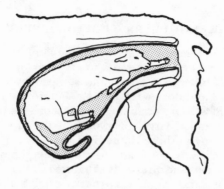

head may drop out of the pelvic girdle, making it difficult to get it started back again.

If the right leg is presented, the ewe should be lying on her right side, so the turned-back leg is uppermost. This would make it easier either to get that backward leg into the right position, or even to help the ewe to lamb even though the leg is not in the normal position.

5. HEAD, WITH BOTH LEGS TURNED BACK

Attach noose onto head (behind ears, and inside mouth). Try to bring one leg down into position, then the other, without pushing the head back any further than necessary. Attach noose of cord onto each leg as you get it out, then pull lamb.

If your hand cannot pass the head to reach the legs, place the ewe with her hind end elevated, which gives you more space. With snare over the lamb's head, push it back until you are able to reach past it and bring the front legs forward, one at a time. Put ewe back in normal reclining position, start head and legs through pelvic arch, and pull gently downward.

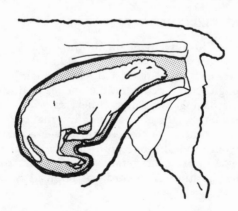

6. BOTH LEGS, WITH HEAD TURNED BACK

Head may be turned back to one side along the lamb's body, or down between its front legs.

If front legs are showing, slip a noose of heavy cord over each front leg, then push lamb back until you can insert lubricated hand and feel for the head position, then bring head forward into its normal position. With noose on legs, you won't lose them. Pulling gently on legs in downward direction, guide the head so that it will pass through the opening of the pelvic cavity at the same time as the feet emerge on the outside.

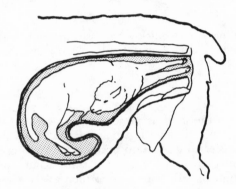

If the head does not come out easily, it is either a large head, or the lamb may be turned on its back (with its back down toward the ewe's stomach). With cords still attached to legs, you may have to push it back again, and gently turn it a half turn, so that its legs are pointed down in normal position, for it will come out easier that way.

If you have a hard time getting a grip on that slippery head to bring it into position, try to get a cord-noose over its lower jaw. Insert your hand with the noose over your fingers, then slip it off onto the chin. Be sure it does not clamp down on any part of the inside of the ewe, and tear her tissues. By pulling on the noose that is over the chin, you can more easily guide the head into position.

7. HIND FEET COMING OUT FIRST

Pull gently, as the lamb often gets stuck when half-way out. When this happens, swing the lamb from side to side while pulling, until ribs are out, then pull out quickly. Wipe off its nose at once so the lamb can breathe. Delay at this point can allow the lamb to suffocate in the mucus that covers the nose.

Sometimes it's easier on the lamb if it is twisted one-half turn, so its back is toward the ewe's stomach, or even rotating it a quarter turn while pulling

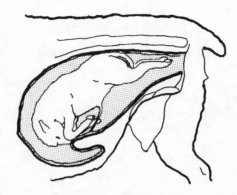

it out. Finish pulling it out quickly because the umbilical cord is pinched once the lamb is half out, and if lamb tries to breathe, it will draw in mucus.

8. BREECH

In the breech birth, the lamb is presented backwards, with its tail toward the pelvic opening, and the hind legs pointed away from the pelvic opening.

It is generally easier to get it into position for coming out with its back feet coming out first, but once you get it started out, speed of delivery is important. The lamb will try to breathe as soon as the navel cord is pinched or broken, so it can suffocate in mucus if things take too long. Wipe off its nose as soon as it pops out.

To deliver, change breech position by positioning the ewe with her hind end somewhat elevated, so that the lamb inside her can be pushed forward in the womb. This will make barely enough space to reach in and slip your hand under the lamb's rear. Take the hind legs, one at a time, flex them, and bring each foot around into the birth canal.

When the legs are protruding, you can pull gently until the rear end appears; then grip both the legs and the hind quarters if possible, and pull downward, not straight out.

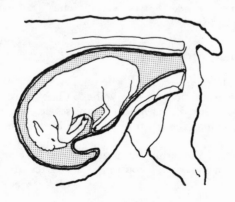

If the ewe is obviously too exhausted to labor any more, try to determine if there is another lamb still inside her. If not, go ahead and give her a penicillin shot, or insert antibiotic uterine bolus, to prevent infection.

However, if you are unable to bring out the legs, and the ewe is making no progress, call the vet.

9. LAMB LYING CROSSWAYS

It sometimes happens that the lamb is lying across the pelvic opening, and only the back will be felt. If you push the lamb back a little, you can feel which direction is which. It can usually be pulled out easier hind feet first, especially if these are closer to the opening. If you *do* push it around to deliver in normal position, the head will have to be pulled around. If it is also upside down, it will need to be turned a half turn to come out easily.

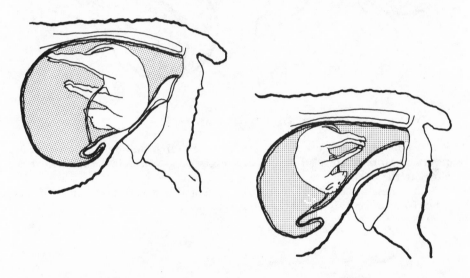

10. ALL FOUR LEGS PRESENTED AT ONCE

If the hind legs are as convenient as the front, choose the hind legs and you won't have to reposition the head. If you choose the front legs, head also must be maneuvered into correct birth position along with the legs. Attach cords to the legs before pushing back to position the head.

11. TWINS COMING OUT TOGETHER

When you have too many feet in the birth canal, try to sort them out, tying strings on the two front legs of the same lamb and tracing the legs back to the body to make sure it is the same lamb, then position the head before pulling. Push the second lamb back a little to give room for delivery of the first one.

Always deliver first the one that requires the least maneuvering—usually the hind-leg one.

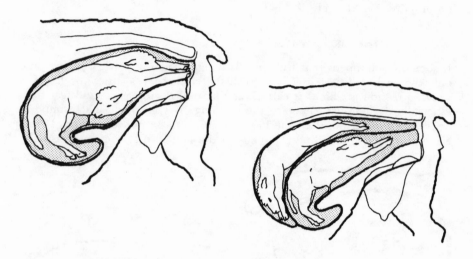

12. TWINS, ONE COMING OUT BACKWARD

With twins coming together, it is often easier to first pull out the one that is reversed. More often, both lambs are reversed, so you pull the lamb that is closer to the opening.

Sometimes, the head of one twin is presented between the forelegs of the other twin, a confusing situation, but very rare.

CHAPTER NINE

CARE OF BABY LAMBS

DOCKING

Before lambs are turned out of the lambing pen, their tails should be docked (removed). This is much easier on the lamb when it is two or three days old, and the tail is still small.

Sheep of most breeds are born with long tails, and these can accumulate large amounts of manure on the wool, attracting flies (and maggots) and interfering with breeding, lambing, and shearing.

There are several ways to remove tails. The docking can be done by cutting (knife not too sharp or wound bleeds more) or with a hot chisel (this sears the wound so less bleeding) or with *Burdizzo* emasculator and knife, or *Elastrator,* which applies small strong rubber rings.

We use the Elastrator because it minimizes shock and eliminates bleeding problems. Apply band at the third joint, which is about 1 inch or 1½ inches from the body. Rubber rings should be stored in a small wide-mouth jar of alcohol to keep them sterile and to disinfect your fingers when you reach in for one. Dip the Elastrator pliers in it, too. The tail falls off in seven to fourteen days, but after three days it can be cut off, on the body side of the band, and the stump dunked in 7 percent iodine or Lugol's solution.

Some people object to the Elastrator docking, believing the lamb is more prone to tetanus. If tetanus is a problem in your location, you should administer tetanus antitoxin (follow manufacturer's directions for dosage) to each lamb at the time of docking and castration, no matter *which* method is used. The tetanus antitoxin is usually sold in 1500 unit vials, and about 150 units is given per lamb. Or tetanus vaccine can be given to the ewes prior to lambing, and that will protect the lambs for about six weeks after birth.

CASTRATION

Castration also can be done early, as soon as the testicles have descended into the scrotum.

Emasculator. If tetanus is known to be a problem, or if you have horses so that it is likely to be present, use an Emasculator for castration, so that there is no wound. This method is also best in late lambing and warm weather, leaving no opening to attract flies. The Emasculator is a pincer instrument that gives bloodless castration by crushing the spermatic cord and arteries when you clamp it onto them like pliers. There is no loss of blood, less pain and setback to the lamb's growth, and no danger of infec-

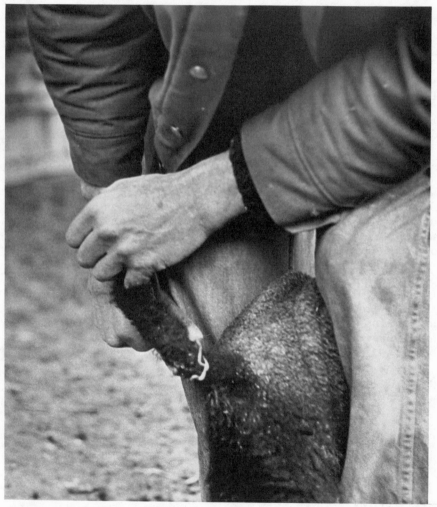

How to apply elastrator band to tail of three-day-old lamb. Special pliers stretch band for easy application.

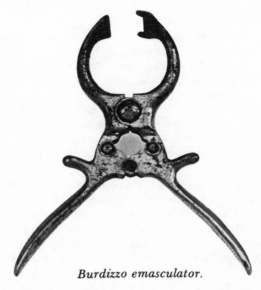

Burdizzo emasculator.

tion. Check to see that testicles have descended into the scrotum, then clamp the Emasculator onto the neck of the scrotum, where it joins the body, on each testicle tube separately.

Because of the high cost of this well-made piece of equipment, you may not want to buy one for use on a few sheep. You might borrow it from a neighbor who has more animals or buy it in partnership with another sheep raiser.

After the Emasculator is used, the testicles will atrophy in about thirty to forty days.

The Emasculator can also be used for docking tails. Keep it disinfected. Push the skin toward the body, and crush between the joints. With a very young lamb, the tail is clamped in the instrument and pulled off. With older lambs, use a knife to cut it off just inside where it is clamped. The main artery is generally so emascerated as to give quick coagulation of blood, and less bleeding than cutting off tail with a knife. Douse the stump with disinfectant and spray with fly repellent if weather is warm.

Elastrator. The Elastrator also can be used for castration, when the lamb is about ten days old and when the testicles have descended into the scrotum. These special pliers stretch the rubber ring so you can pull the scrotum through it, being sure both testicles are down. When the pliers are removed, the ring tightens where it is applied, around the neck of the scrotum where it attaches to the body, cutting off the blood supply so the testicles wither usually within twenty to thirty days. There is no internal hemorrhage, or shock, and risk of infection is slight. If you have problems with infection, douse the band with iodine after about a week. In hot weather, you can spray it with fly repellent.

Is castration necessary? There are reasons for not castrating—*if* you will be marketing the animal for meat at five or six months of age, or are thinking of keeping or selling it for a breeding ram.

While castrated lambs grow faster than ewe lambs, the uncastrated males will outgrow both of them, and the meat will be leaner. So, if you have early lambs and plan on selling the rams for meat at five to six months old (before breeding season), you can omit the castration. However, a packing house may penalize you $1 per animal or one cent per pound, for not castrating, if that is your market. If you intend to keep the ram longer than six months before slaughtering, castration is desirable.

Cryptorchid or short scrotum. There is still another approach, where the rubber Elastrator ring is used on the scrotum, but the testes are pushed back up into the body cavity. This sterilizes the animal, probably from the body heat, and while the male hormones are still present to increase weight gain with more lean meat, the animal shows little or no sex activity. This method is used at about four weeks of age, and the animal is called a *cryptorchid* (meaning "hidden testicles"). Extensive tests in Australia have shown such animals gain weight faster, get to market size faster, and have less fat and more lean than either castrated or uncastrated males.

Scientists say that for the large sheep operations, the laser beam may someday be used for both castration and docking. It will be painless, not stress the animal, and cauterize as it cuts. But this is still a long way off, and will be economical only on a grand scale.

Methods used to induce cryptorchidism. (Shepherd *magazine, Dec. 1973.)*

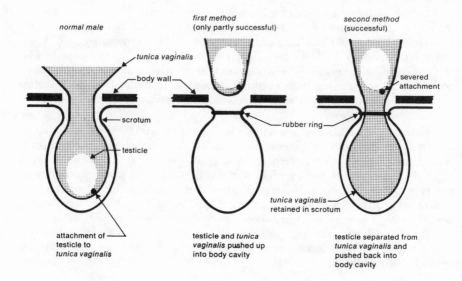

FEEDING LAMBS AND EWES

A ewe with twins (or triplets) cannot consume enough grass to support herself and give milk for them to grow, so she will need hay and grain until they are weaned. Even ewes with single lambs should have supplemental feed, unless you have lambed late and have a lot of pasture.

A ewe with a single lamb should have approximately 1 pound of grain a day, while a ewe with twins should get 1½ to 2 pounds a day, plus some hay. Lambs from heavy milking ewes can gain up to 70 percent more during the nursing period than those from poor milkers. Lambs from good milkers will double their birth weight in two weeks.

In addition to their mother's milk, and the grass they start to nibble at about ten days old, the growing lambs need grain and hay in their own feeder, called a "creep." Start creep feeding early, as young lambs will investigate and use the creep more quickly than older ones, and it helps to establish their rumen function.

CREEP FEEDING

The creep is an enclosed space lambs can enter and eat all they want, but ewes cannot enter because of the size (8"–9") of the doors and openings. The creep should be sheltered, with good fresh water provided daily, and be well bedded with clean hay or straw. The heavy stems of alfalfa, left uneaten in the ewes' hay rack, are good creep bedding. If the creep is in the barn, it should be well lighted, because lambs prefer it that way and eat better. Hanging a reflector lamp four or five feet above it will attract the lambs. They can start using the creep when they are about two weeks old.

Feed stores sell special "lamb creep feed," but if it is not available, you can start them off with a mixture of crushed whole grains, plus some of whatever their mothers eat. They always prefer the same grain as the ewes eat, so sprinkle some of it on top of the creep mixture, if you buy that for them.

USDA recommends, if no creep feed ration is available, a mixture of the following percentages: 60 corn, 20 oats, 10 bran and 10 soybean meal, with about 1 percent bonemeal and 1 percent mineralized salt added. This mixture can be coarse ground at first, then fed whole later.

Grain and leafy hay are best given fresh twice daily, with uneaten portions fed to the ewes. To get the lambs to eat, the food must be attractive to them.

FORWARD CREEP GRAZING

If you have several pastures and rotate their use, you can give the growing lambs the benefit of the best grass by allowing them into fresh pasture ahead of the ewes, through creep-type openings in the fence. This will save creep grain, the lambs will grow faster, and they will have fewer internal parasites.

FACING OR LAMB COAT

Our northwest area has a moderate climate, but a long rainy season. Sheep don't mind a light rain. It usually takes a heavy rain, plus wind, to get them to go into the barn. This can create a problem when there are small lambs. If you keep the small ones penned for the first three days, as we do, they are quite hardy after that.

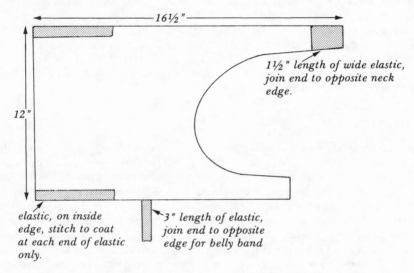

←————————— 16½ " —————————→

12 "

1½ " length of wide elastic, join end to opposite neck edge.

elastic, on inside edge, stitch to coat at each end of elastic only.

3 " length of elastic, join end to opposite edge for belly band

Here's a lamb coat for very cold weather, made from canvas or duck.

Then bad weather can be met in two ways. One is by *facing,* or trimming the wool from the sheep's face so she will seek shelter, and take her lamb. Facing is called "wigging" in Australia and New Zealand. The other is to use a canvas lamb coat, although some ewes find them disturbing. There is a new plastic lamb coat available, made in perforated rolls, and individual coats are torn off the roll as needed. These are useful when lambing outdoors, with no shelter available for the newborn lambs.

CHAPTER TEN

ORPHAN LAMBS

LAMBS & EWES

When a ewe has twins and rejects one, you have three options: Persuade her to accept it; graft it on another ewe who has lost a lamb or only has a single; or bottle-feed the lamb. (More on bottle feedings later.)

PERSUADING THE EWE

If a ewe has a single lamb which she rejects, you have a more frustrating situation, for in addition to a hungry lamb you have an increasingly uncomfortable mother. In this event, it is urgent that the lamb get some colostrum in its stomach, so you roll the ewe on her side and put the lamb's nose against her teat. In most cases the lamb is hungry and very cooperative. This first feeding gives you a little time to arrange a forced acceptance of the lamb.

Should a ewe reject the lamb *after* it starts to nurse, not before, check its teeth. They may be painfully sharp. A little filing with an emery board can remedy the situation. Don't file too much or the lamb's teeth will be sore and it won't nurse, which will put you right back where you were. Apply *Bag Balm* to the ewe's teats, if they are lacerated by teeth. Keep her tied where the lamb can nurse until she accepts it.

Once a ewe rejects a lamb for *any* reason, it is hard to fool her into accepting it. Since she identifies her lamb primarily by smelling its rear end, you can try daubing her nose, and its rump with vanilla; *U-Lam* aerosol or *Mother Up*, both made for this purpose; room deodorant, non-scented; or a squirt of her own milk.

If the mother is a yearling, high strung and not very tame, a tranquilizer can also work wonders.

There is also an old-timer's method of tying a dog near her pen. Its presence is supposed to foster the mothering instinct. We find this sometimes makes the ewe so fierce that she will butt the lamb if she can't reach the dog.

Another solution is to keep the mother penned in such a way that she cannot hurt the lamb, and it can nurse regularly in safety. This sometimes can be done in the lambing pen by tying the ewe. If necessary, her hind legs can be restrained temporarily so she can't keep moving them and preventing the lamb from nursing. Any system of tying will necessitate frequent supervision, in case she becomes entangled and chokes.

When the ewe is acting up, the lamb is left without its mother's guidance and encouragement. Hence you will need to help the lamb nurse by holding the ewe, and pushing the lamb to the right place.

If the ewe still kicks at it and will not hold still for nursing even though the lamb is trying, she will have to be restrained. A ewe stanchion can be used. A less elaborate one can be improvised in the corner of the lambing pen. If you use a board along her side, it should be high enough for the lamb to approach and nurse, yet leave enough room for the mother to be able to lie down when she wants to. She will need hay in front of her, and drinking water either there or offered to her often. Put molasses in it, as you would for any ewe who has just lambed. It may take from one to three days before she is resigned to accept the lamb.

The most typical situation is the birth of twins, and the rejection of *one* of the twins. Try spraying the rear ends of *both* of the lambs with a confusing scent.

If a ewe starts showing any hostility *at all* toward one of her twins (either by acting suspicious of one, or talking with soft baby talk to one and a grumpy sound to the other), we don't wait until she starts butting it, but take positive action right away. The system we have found most successful is to tie her up. The sooner you stop her from comparing the smell of the two lambs, the sooner she will accept the reject.

If you don't have a ewe stanchion, tie her in a corner so she can't get her head around to sniff the lambs. She has to have enough rope to lie down, so if you place a bale of hay against the wall she is facing, with the end of the bale against her lower shoulder, she will not be able to turn her head to where the lambs are nursing. We put a few rocks or some flat pieces of scrap metal under her, something that won't make her too uncomfortable but will discourage the lambs from lying down under her.

She should need to be tied no more than two days. Be sure she gets water occasionally, for with a make-do arrangement it is difficult to leave it in front of her.

Some suggest that you tie up the ewe and leave the reject lamb with her, but take away her favorite lamb, bringing it back only to nurse. We have not found that necessary, and I would hesitate to do it because I'd be so frustrated if she accepted the reject, and decided she didn't want the one I had taken away.

In the meantime, if another ewe is in labor and you think she may deliver only one lamb, you might choose to graft on the reject, for she may be more willing than the ewe who is all geared up to rejecting something.

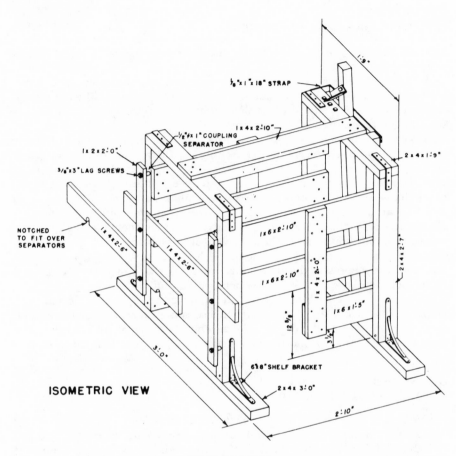

ISOMETRIC VIEW

LIST OF MATERIALS

LUMBER:	MISCELLANEOUS:	
2 pcs. — 2 x 4 x 10'-0''	6	— ½'' x 1'' pipe couplings
1 pc. — 2 x 2 x 7'-0''	4	— 6'' x 8'' shelf brackets
2 pcs. — 1 x 6 x 8'-0''	1	— ⅛'' x 1'' x 4'-0'' steel strap
1 pc. — 1 x 4 x 12'-0''	1	— 1½'' x ⅛'' x 5'' steel strap
1 pc. — 1 x 2 x 4'-0''	4	— ¼'' x 2'' lag screws and washers
	10	— ⅜'' x 3'' lag screws and washers
	48	— 1¼'', No. 9 wood screws
	¾ lb.	— 6d common nails
	¼ lb.	— 12d common nails

This ewe stanchion, USDA Plan 5912, is widely used whenever ewe refuses to claim lamb. The corners of the neck boards, the back corners of the crate, and the rear restraining boards should be rounded to reduce chances of injury to the ewe and wearing of the wool.

GRAFTING ORPHAN ON A DIFFERENT EWE

Get a bucket of warm water ready, have the rejected lamb nearby, and watch the lambing.

As the ewe delivers her own lamb, dunk the waiting reject into warm water up to its head, then rub the two lambs together, especially the tops of the head and rear ends. Present them both to the ewe's nose, and usually she will lick them and claim them both.

However, if she delivers twins, you'll have to take the reject back. Dry the lamb out, and keep trying to get its mother to take it (or bottle-feed it yourself).

If the substitute mother does appear to accept it and the lamb is much older than her own newborn, hobble the orphan's legs so it doesn't get up and run around too much at first. Let the newborn lamb have the first chance to nurse. If your orphan is a few days old, it doesn't really need the colostrum, and should not get too much of it at one time.

To do this trick, the orphan should be less than a week old, as an older one would surely cheat the new lamb out of its share of the milk. In any event, both lambs will have to be supervised carefully.

When the ewe is lambing, try to catch the water bag as it is ejected, and empty it into a bucket. Then you can use this instead of warm water to dunk the orphan. Don't neglect the newborn during all of this. Its nose must be licked off by its mother or wiped off by you, so that it can breathe.

GIVING ORPHAN TO EWE
WHO HAS LOST HER LAMB

If you find a ewe who has delivered a dead lamb, and you have a young orphan who needs a mother, dunk the lamb in warm water containing a little bit of salt and some molasses. Dip your hand in the warm water and wet its head. By the time she licks off the salt and molasses, she usually has adopted the lamb.

In all this talk about grafting an orphan onto a ewe, I've not mentioned the old way of the "dead lamb's skin." In that method, if a lamb were born dead, or died after birth, it was skinned and the skin fastened like a coat over the orphan. Skinning a dead lamb is not simple unless you already are adept at it. The process is messy and unsanitary, since you may not know why the lamb is dead, and could be transferring germs and disease.

Another less messy method is to rub a damp towel over the dead lamb, then rub the towel on the orphan. Before doing this, wash the orphan with warm water, giving special attention to washing the rear end which is the first place the ewe checks in determining whether the lamb is her own.

I remember a postage stamp issued a few years ago, showing a ewe with a lamb. She appeared to be sniffing its head. Sheepraisers laughed, as it was a very *un*-typical end for her to be sniffing.

The fewer sheep you raise, the less chance there is that another ewe will

Targhee ewe with newborn lamb. (Mt. Haggin Livestock Inc., Anaconda, MT)

be lambing about the time you need a substitute mother. So if its mother has died, has no milk, has been incapacitated by pregnancy disease or calcium deficiency, or completely refuses to accept her baby, you have a bottle lamb.

BOTTLE LAMB

This is one of the greatest pleasures (and biggest headaches) of sheep raising.

The lamb's first need is having its nose mucus wiped off so it can breathe. Even if the ewe is weakened by a hard labor, and/or has no milk, she should be allowed to clean the lamb as much as she will; if unable to nurse, she will still "claim" it, and even as a bottle lamb, it can stay with her. If the ewe does not lick off its nose, you wipe it off, then dry the lamb and put iodine on its navel.

Now it needs some real colostrum (ewe's first milk) if possible, either from its mother who may have rejected it or is too weak to stand up (roll her over and help the lamb), or from another newly-lambed ewe, or some defrosted if you have it frozen. Cow or goat colostrum are the next best substitutes for ewe colostrum.

NEWBORN LAMB MILK FORMULA*

If no colostrum is available, use this newborn lamb formula, which can be fed for the first two days:

> 26 ounces milk (½ canned milk, scant ½ water)
> 1 Tbsp. castor oil or cod liver oil
> 1 Tbsp. glucose or sugar
> 1 beaten egg yolk

Mix well, and give 1½ to 2 ounces at a time the first day allowing about two hours between feedings. Use a baby bottle and enlarge the nipple hole to about the size of the head of a pin. (Lamb nipples are larger; use when lamb is older.)

On the second day, increase the feedings of the formula to 3 ounces at a time (4 ounces for a large, hungry lamb) two hours apart. On the third day the formula can be made without the egg yolk and sugar, and oil can be reduced to 1 teaspoon into the 26 ounces of milk. One child's vitamin tablet, crushed and dissolved in the milk, is a good idea on the third day.

In this formula, goat's milk is better than diluted canned milk. If you have plenty of goat's milk, you won't have to buy the lamb milk replacer that you would be gradually changing over to after the first three days on the first formula. Do not overfeed at any time or the lamb will get sick. It's better to underfeed.

A bottle lamb is more subject to infections than a lamb on the ewe's milk, so keep bottles and nipples clean.

If the droppings get runny (diarrhea is called scours), use *Pepto Bismol* immediately. Give several teaspoons for a lamb a week old or younger; give more for a larger lamb, and cut back on the amount of milk given, diluting it more. This condition is often the result of overfeeding.

Keep milk refrigerated, warming it at feeding time, and keep milk containers clean, to prevent the kinds of scours caused by bacteria.

LAMB MILK REPLACER

After using newborn lamb formula for the first two days, gradually change over to "milk replacer" powder, bought at the feed store. Get only *lamb* milk replacer, made especially for lambs. The calf milk replacer is too low in fat and protein. Don't use it.

The lamb milk replacer, sold under several brand names, comes in 25-

---

*Bottle lambs can be raised successfully even when they do not get colostrum with its important antibodies that help the lamb withstand disease. Anyone who likes interesting do-it-yourself projects might want to make up another source of antibodies for use if needed. See end of this chapter on blood serum antibodies.

and 50-pound bags; if the store doesn't have it, ask for a special order. We use *Land-O-Lakes* (the kind for lambs; they also make another for calves), but there are other brands, such as *Ewelac* and *Lama*.

CONTENTS OF MILK

A comparison between the contents of cow's milk and sheep's milk will show the difference there will be in the replacer milk made for their offspring:

		Grams of Protein	Grams of Fat	Grams of Calcium
Ewe milk	1 cup	11.0	13.0	0.413
Cow milk	1 cup	7.0	7.8	0.236

From *Table of Food Values* by Alice V. Bradley.

Ewe's milk has almost twice as much fat as cow's milk, and the lamb milk replacer has about twice the fat content, on a solids basis, as calf milk replacer. A newborn lamb is a small creature with a high ratio of body surface compared to its heat-generating capacity (more so than a calf). The fat provides this heat generation. Nature provided ewe milk with high fat content for a good reason.

Carl Hirschinger of the University of Wisconsin says to look for a lamb milk replacer that contains 30 percent fat and 24 percent protein, on a dry matter basis, and no more than 25 percent lactose (milk sugar). High lactose levels can cause diarrhea and bloat. Whey products are high in lactose, and should be delactosed, or used sparingly in the replacer. Hirschinger recommends diluting replacer to a minimum of 15–20 percent dry matter for the first week.

Most lamb milk replacers are medicated for prevention of many common lamb ailments, and fortified with Vitamin A, which is very high in the natural ewe's milk, as well as with Vitamin D and Vitamin E, which is useful in the prevention of stiff lamb disease. The milk replacers also have necessary minerals.

We mix the replacer powder in a blender, make enough for two days, and store it in the refrigerator. We mix ours double strength, using only half as much water as the recipe calls for, and then dilute it with an equal amount of hot water for bottle feeding.

One suggestion I would make for mixing milk replacer: Substitute *limewater* (see recipe below) for about one-half of the water in the concentrated, double-strength formula.

To make limewater, use ordinary slaked lime purchased from a feed store. If you have trouble finding the slaked lime, buy it as calcium hydroxide in the drug store. You need only a very small bottle of it. The 0.4-ounce bottle will last for a lamb's three months of feedings. Some drug stores sell the plain limewater already made up (this is *not* the juice of limes) because it is used in some baby formulas.

LIMEWATER RECIPE

1 tsp. slaked lime
1 gallon of water

Add the lime to the water. Shake it several times during the day, then let it stand until it is clear. Drain off the clear liquid and use as ¼ of the water in the milk replacer mixture. (Or, you can make up half of the limewater recipe at a time with ½ tsp. lime to 2 quarts of water.)

Limewater is sedative, antacid, astringent, makes milk more easily digested, and reduces the tendency to scours.

A slow-flowing nipple is best, for then the lamb cannot gulp its milk and choke or get sick. A hole about the size of a pin head is about right for the first couple of weeks and can be enlarged just a little as needed. By the time the lamb is a month old, you may want to change to a lamb nipple, which fits over the top of a soft-drink bottle.

As the lamb gets older, it drinks more and gets correspondingly louder at mealtime, bouncing its bottle like a punching bag and making a real pest of itself. Sooner or later the inevitable happens; it takes a mighty swig at the bottle and pulls off the nipple, splashing everyone with milk and creating a threat of disaster — the possibility that it will not only chew but swallow the removed nipple before you can take it away.

BLOATED BOTTLE LAMB

This is an infrequent situation, but it can happen if the lamb is overfed or if it drinks too fast (nipple hole too large). Cut back on the amount of milk

The lambs come and get it at this New Zealand experiment in raising large numbers of lambs on bottles. Note each lamb has its own bottle. (D.H.B. MacQueen, photographer, Raukura Agricultural Research Center)

Finnsheep sextuplets. Four of these were raised as bottle lambs. (USDA, Beltsville)

being given, and give one small feeding of two ounces of milk containing about one tablespoon (for lamb under one month) or two tablespoons (for lamb over one month) of human antacid medication with simethacone. This is similar to a veterinary bloat remedy called *Bloat Guard,* containing methyl silicone. If the lamb will not take it in the bottle, give with spoon, carefully.

SUGGESTED FEEDING SCHEDULE
FOR ORPHAN LAMB

Age	Amount
1–2 days	2–3 ounces, 6 times a day, approx. (with colostrum or newborn lamb milk formula)
3–4	3–5 ounces, 6 times a day (changing over to lamb milk replacer)
5–14	4–6 ounces, 4 times a day
15–21	6–8 ounces, 4 times a day, and start with leafy alfalfa hay and crushed grain, or pelleted creep feed*
22–35	Slowly change to 1 pint, given 3 times a day. After lamb is three months old, feed whole grain, and alfalfa, or pelleted alfalfa containing about 25 percent grain.

*See creep feeding section, Chapter 9.

If you raise Finnsheep that have many multiple births or a flock large enough to have quite a few orphans, there are several commercially-made, multiple-nipple arrangements for feeding them, such as *Lam-Bar* and *Lambsaver.* The lambs first should be taught to nurse from a bottle on warmed milk-formula, then changed to the multiple-nipple feeder.

With this system, milk is always before the lambs, and they can suck it out as they want it. The milk is usually fed cold to reduce the chance of overeating, and to reduce bacterial contamination when it is left standing all day. This lamb milk feeder should be cleaned and disinfected and supplied with fresh milk daily. For such use, the replacer should be one that stays in suspension well.

D.T. Torell, at the University of California's Hopland Experiment Station, did tests showing that the addition of one milliliter of Formalin (37 percent solution of formaldehyde in water, available in small quantities from most pharmacists) to each gallon of liquid milk replacer being fed at barn temperature, would keep the milk free of bacteria for several days without having any adverse effect on the lambs. While it is still preferable to have fresh milk almost daily, the tiny amount of Formalin would eliminate the chore of such careful disinfecting of the feeding equipment. The one milliliter of Formalin is the same as one cubic centimeter on your hypodermic syringe. You also might measure it with a one cubic centimeter insulin syringe.

Here are illustrations of the Hopland Lamb Nursing System, developed by Mr. Torell. It is a gallon jug suspended upside down, with a No. 6 rubber stopper from which a ¼-inch diameter plastic tube protrudes. The tube then leads to another No. 6 stopper holding a nipple which is inserted

This lamb feeding arrangement was designed by D. T. Torell of the Hopland Experiment Station. The Lam-Bar *nipples used with it can be purchased from Sheepman Supply (see Appendix).*

through the board to the lamb area or pen. The nipple is available for nursing at all times.

The hole in the board for the nipple must be located one to two inches above the level of the jug neck. A strip of inner tube or other strong elastic can be tied on each side to screw eyes, to hold the jug in place. A similar arrangement can be used to keep the stopper and nipple firmly in the hole while the lamb nurses. One of these Hopland Nursing Systems will feed three or four lambs, using lamb milk replacer fed at barn temperature.

HOW TO HARVEST ANTIBODIES
FROM BLOOD SERUM
FOR ORPHANS

When a healthy sheep is slaughtered for locker meat, collect the blood in a well-sterilized container. Let the blood stand at room temperature until it has clotted and the clear yellow fluid has come to the top. Pour off this fluid (which is the serum) into sterilized containers, filling two-ounce containers about one-third full. If kept frozen solid, the serum will keep up to two years for use when no colostrum is available for a lamb.

An orphan lamb can be given a subcutaneous injection of one of these jars (less than one ounce), under the skin of the rear armpit or the neck. Serum must be thawed at room temperature, as use of heat will destroy its usefulness.

You can see that freezing extra colostrum, for use when needed, is the easier way to provide the antibodies needed by a newborn lamb.

CHAPTER ELEVEN

LAMB PROBLEMS

WEAK LAMB AT BIRTH

A lamb that has a difficult birth, and is very weak, may be born with some fluid in its respiratory tract. If it gurgles with the first few breaths, dry off its nose, then hold it up by the rear legs so that the fluid may have a chance to drain out.

If the heart is beating, but the lamb does not start to breathe, hold it by its hind legs and swing it in a circle. Then give gentle artificial respiration by breathing into its mouth. Don't blow as though blowing up a balloon; its lungs are quite small. If the lamb is warm, heart beating but not breathing, and your attempts at artificial respiration do not help, sometimes a cold water shock treatment will do it. Dunk the lamb in cold water, even a drinking trough, and the shock may cause the lamb to gasp and start to breathe. Then, warm it up, and get it to nurse.

You should avoid excessive heating and the unnecessary use of the heat lamp, for this leaves the lamb prone to pneumonia. If a newborn lamb is so cold from exposure after birth that its inner mouth is cold, there is no time to warm the lamb with a heat lamp. It is much quicker to revive it in hot water. The water should be comfortable to the touch when the lamb is immersed, then heated to about 100 degrees over a period of five to ten minutes. Move its legs around in the water to increase its circulation.

Until recently lambs were immersed in hot water, but some deaths attributed to chilling were found to be due to shock, which does not happen when the water is warmed more gradually. Keep the lamb in the water until its mouth is warm, then rub the lamb dry, give warm milk if it will suck, and wrap in a blanket until it seems quite dry. Some veterinarians suggest injections of $\frac{1}{4}$ cc of Vitamin A (or $\frac{1}{2}$ cc of Vitamin ADE) to prevent pneumonia.

If you have no electricity in the lamb pen, use a hot water bottle, or a plastic jug filled with hot water, to warm a lamb until it is dried off. This would not be sufficient for extremely chilled lambs. When using a hot water bottle (not hot enough to burn), apply the heat first to the belly where it is most needed.

If a lamb a few hours old is crying continuously or has a cold mouth, it is not nursing. The ewe may not have milk (or you may have forgotten to strip her teats to unplug her), or the lamb may not have found her udder. Check to determine the problem.

If a weak lamb has not been able to stand up to try to nurse within a half-hour, help by holding it up to the ewe if she will stand still, or put the ewe down and hold the lamb to nurse. Use this same procedure for a stronger lamb, if it has not located the right place and begun to nurse within one hour after birth. We seldom wait that long. Just roll the ewe on her side, check that she is unplugged, and hold the lamb to her udder, placing the teat in the lamb's mouth if necessary.

For a very weak lamb you may have to give it the first feeding from a baby bottle with the nipple hole enlarged to about the size of a pin head. Use about two ounces of the ewe's colostrum, warm, to give it strength. Do not force the lamb, if it has no sucking impulse, or the milk will go into its lungs and cause death. If it has no suck, try the dextrose injection described below (this is easy), and wait for a half-hour to see if this gives it energy and the desire to suck. If not, then try the stomach tube feeding method described later in this chapter.

If the ewe has no milk, and you have no frozen colostrum or another newly lambed ewe to swipe it from, see the colostrum formula in Chapter 10 and proceed as for an orphan lamb, but keep it with its mother if she is attentive, for the lamb will be happier and healthier with a sheep-mama to follow around.

DEXTROSE INJECTION REVIVAL, FOR VERY WEAK NEWBORN LAMB

Do not attempt to feed a very weak lamb with a bottle, if it cannot suck. If it is too weak to suck, you can give it a real shot of energy by a dextrose injection, using

10 cc of 50 percent dextrose in saline, or glucose in saline, warmed to body temperature

or

50 cc of 5 percent dextrose in saline, warmed to body temperature.

Warm either of these solutions to body temperature, and divide the dosage so that it may be injected subcutaneously (beneath skin) in the loose unwooled skin under each rear armpit of the leg. The 10 cc of 50 percent can go into two locations, while the 50 cc of 5 percent should be divided and injected under all four legs. You should sterilize the injection sites and the top of the bottle with alcohol, and also sterilize the syringe and the needle (#18 or #19 needle). Inject slantwise under the loose skin of the arm-pit. As you withdraw the needle, wipe the skin again with alcohol-wetted cotton, and massage the area to distribute the dextrose under the skin. One of the dextrose strengths usually can be obtained from a local pharmacist, or any hospital. Or, you can order a preparation by mail that is called *Caldex M.P.*, which is 23 percent calcium gluconate, 25 percent dextrose monohydrate, and magnesium chloride. Use 2 cc to 5 cc, divided into two injection sites, under the skin of the rear armpits. Caldex M.P. also is used to treat pregnant ewes, an alternate or in addition to propylene glycol given by mouth, for toxemia.

A dextrose injection could give the lamb the strength to nurse within 15 minutes to a half-hour. If you see no improvement, try stomach tube feeding, using the following instructions.

WEAK NEWBORN LAMB—STOMACH TUBE EMERGENCY FEEDING

The Sheepman Supply Company (see Appendix) has a device called a *lambsaver* that you can order by mail, for the stomach-tube feeding of a severely weak lamb that has no sucking impulse. If you need one quickly, there is no time to order, so get a male catheter tube from the drug store, and use it with a rubber ear-syringe of 4-ounce capacity, or a 10 cc hypodermic syringe, for a direct feeding into the lamb's stomach. The tube should be about 14–16 inches long. Before inserting to inject the milk, disconnect the tube from the milk-filled syringe, so that you can determine that the tube is actually in the stomach and not in the lungs. An injection into the lungs would kill the lamb.

One arrangement for emergency stomach tube feeding.

When the tube is in what you think is the correct position, hold a wet finger at the protruding end of the inserted tube. If the finger feels cool from moving air, the tube is in the lungs instead of the stomach, so remove it and try again. It is easier for two people to do this, but it is possible with one person, if the syringe is already filled with warm colostrum (or warmed canned milk, two to three ounces not diluted for this feeding only) and within reach. This is the procedure for one person:

Position of lamb: On a table, with its feet toward you. Hold the lamb's body with your left forearm, making a straight line between the lamb's head and neck and back. Use fingers of left hand to open the lamb's mouth to insert the tube, which should be sterile and warm, if possible.

Insert the tube slowly over the lamb's tongue, back into its throat, giving it time to swallow, and push the tube down its neck and into the stomach, having checked the catheter tube length previously, so you know about how much of it should stick out, when the end is in its stomach area. The average insertion distance is 11 or 12 inches. You cannot insert it too far, but it is important to insert it far enough.

When you have confirmed its correct position with the wet-finger test, insert the end of the catheter tube into the syringe filled with warmed milk, and slowly squeeze the milk into the lamb's stomach. Withdraw the tube quickly, so it will not drip into the lungs on the way out. Helen Lund, of *Shepherd* magazine, told me the details of this emergency treatment in a letter in 1966. They had used the procedure only a few times then, and cautiously. We have used it only when we felt that without it the lamb would surely die, and have found it successful, but frightening to do.

ENTROPION
(Inverted Eyelids)

Frequently when a lamb is born, the lower (or upper) eyelid, or both, may be rolled in. When this happens, the eyelashes chafe the eyeball, causing the eye to water constantly, inviting infection, and even blindness. The tendency is hereditary, but more prevalent in woolly faced breeds. Do not keep such a lamb for breeding. Mark it with an ear tag or notch, for slaughter.

Inspect each lamb at birth, so the condition is found at once and corrected. To fix it, roll the eyelid (or eyelids) outward and hold in proper position by clip, or sewing. The use of two little metal "surgical clips" is easier than stitching. They can be clipped into place with forceps or small pliers.

If you want to try sewing them down, use white cotton thread and a sharp needle. Roll the eyelid out, put the needle through a small piece of skin, and sew it down (the upper eyelid is sewn to the forehead, and the lower eyelid to the jaw). In a few days the eyelids will have conformed to a normal position and stitches can be removed. Use a mild antiseptic in stitching or applying clips.

Entropion, *showing lower eyelid turned in. Eyelashes will irritate the eyeball. Pocket behind eyelid becomes infected if condition is not remedied.*

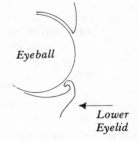

Eyeball

Lower
Eyelid

If you don't want to stitch and don't have clips, pieces of Scotch or adhesive tape may suffice. They may have to be reapplied several times, to hold the eyelid in place for a few days.

PNEUMONIA IN LAMBS

Pneumonia is estimated to be responsible for as many as 15 percent of the lamb deaths. It can be caused by exposure to drafts in cold damp quarters, by overheating as with the overuse of heat lamps, followed by exposure to cold, and also by transmission of germs from the ewe to the lamb.

To prevent pneumonia, lambing pens in open-sided barns can have burlap bags or other material draped over the side facing the wind, to prevent drafts. Use heat lamps no more than is really necessary.

If pneumonia is a real problem in your young lambs, you can decrease the incidence of germs being passed from the ewes to the lambs by using antibiotics on pregnant ewes, according to research scientists at Ohio State University. They suggest use of one pound of sulfamethazine powder (or a gallon of 12.5 percent solution) in 120 gallons of water, offered to the ewes two weeks before lambing. Give the ewes treated water for five days, remove it for three days, and offer treated water for three days. Repeat the three days of treatment at three-day intervals until she has her lamb. Each ewe may drink from 1 to 1½ gallons per day.

MECHANICAL PNEUMONIA

Mechanical pneumonia is caused by fluid or milk entering the lungs of the lamb. An abnormal birth position, which pinches the umbilical cord before birth is complete, can cause the lamb to inhale fluid, resulting in mechanical pneumonia. Also, the forced bottle feeding of a lamb who is too weak to suck sometimes causes liquid to enter the lungs. There is no known cure for mechanical pneumonia.

SCOURS (Diarrhea)
IN NURSING LAMBS

Diarrhea in newborn lambs (called "scours") has many causes. A white kind of scours is usually from too much milk consumption — a strong lamb sucking a mother who has an excess of milk. Milk out the mother somewhat to reduce the amount of milk available, and give the lamb a couple of teaspoons of Pepto Bismol, to firm up the droppings and form a protective coating in the stomach. If scours continue for more than a day, the lamb will need preventive treatment for dehydration, such as a subcutaneous (under the skin) injection of 50 cc of 5 percent dextrose in saline, warmed to body temperature and injected in four armpits. For injection instruction, see Chapter 16.

Bright yellow runny scours is usually from an internal infection caused by unsanitary conditions, such as lambing in a mucky place, soiled bedding in the lambing pen, or ewe not crotched so the lamb sucks on dirty wool tags.

Sulfa antibiotic preparations are helpful for most of these bacterial infections. We like one that has a stomach coating action, along with sulfa, called "Bacterial Scour Treatment," from the Omaha Vaccine Co. (see Suppliers, Appendix A). In dosing a day-old lamb, we give about three teaspoons, orally, administered carefully so the lamb does not choke, with the dose going into the lungs. This medication also has pectin, bismuth, and kaolin to soothe the stomach and firm up the feces. There are other scour preparations that contain vitamins in addition to sulfa or neomycin.

For scours in a bottle lamb, discontinue milk feeding at once. For one day, feed either limewater (see recipe in Chapter 10) or the formula given below, at rate of two ounces about two or three hours apart. The lamb needs the liquid to prevent dehydration. Bottle-lamb scours is usually the result of over-feeding the lamb, or too large a feeding instead of smaller feedings given often. In a few cases, it is the feeding of bacteria in unwashed bottle, or milk-replacer left too long at room temperature.

For bottle lamb with scours, replace milk with the following mixture for a few feedings:

1 quart water
2 ounces dextrose
½ tsp. salt
¼ tsp. bicarbonate of soda

Give this only for one day, or part of the day, until the diarrhea ceases, then return to milk feedings, given in smaller quantities than before.

In addition to the bottle of this substitute milk, the bacterial scour treatment should also be given. You may be able to get the lamb to take some of it in a bottle, dosage from about three teaspoons for a one- to three-day-old lamb, or up to three tablespoons for a two- or three-week-old lamb.

CONSTIPATION IN LAMBS

A constipated lamb usually stands rather humped up, looking uncomfortable, with no sign of droppings, or only a few very hard ones. Sometimes the lamb will grind its teeth, if the condition continues, and will go into convulsions and die unless medicated. Administer two tablespoons mineral oil or one tablespoon castor oil, for a very small lamb (under two weeks old). For one as old as two months, give ¼ to ½ cup mineral oil. Sometimes dosage will have to be repeated.

"Pinning" is an external kind of stoppage that is fairly common in very young lambs, usually under a week old. The feces collect and dry into a mass under the tail, gluing it down and plugging up the lamb. If not noticed and corrected, the lamb may die. Clean off the mass with a damp rag or a paper

Healthy little lambs.

towel, trimming off some of the wool if necessary; disinfect the area if irritated, and oil it lightly to prevent further sticking. Check the lamb frequently. This is another good reason to keep mother and lamb in the pen for the first three days, so you can easily inspect for this and other problems.

NAVEL ILL

This is a name used to describe infections from any number of organisms that gain entrance to the lamb's body through the umbilical cord, shortly after birth. They develop into serious illnesses, usually within a few days.

By treating the umbilical site with iodine as soon as possible after birth, and seeing that the lamb nurses its mother within the first hour (because the colostrum contains antibodies for many of the germs in its environment), you can minimize the danger of navel ill. Clean bedding in the lamb pen will lessen the chance of infection.

The acute form of navel ill causes a rise in temperature, no inclination to suck, and usually a thickening can be felt around the navel. Death follows quickly.

Tetanus is one of the serious diseases caused by a bacillus that can enter through the cord. If you keep horses or are in a known tetanus area, treat the navel with extra strong veterinary iodine, or Lugol's solution from the pharmacy. As an added precaution, treat again the next day. A more certain protection against tetanus is obtained by vaccinating the ewes in the last two months of pregnancy (two separate shots a few weeks apart) with Clostroid DT from your veterinarian. This protects the ewe and passes protection along to the lamb in the colostrum. The vaccine is for both tetanus and enterotoxemia, another of the diseases that can strike lambs.

Since navel ill can be caused by various bacteria, it takes a veterinary diagnosis to determine the specific cause and thus administer the proper antibiotics.

URINARY CALCULI
(Stones, "Water Belly")

A problem of growing ram lambs over one month old, castrated or not, is that the salts they normally excrete in the urine can form stones and these may lodge in the kidney, bladder or urethra.

Symptoms. The lamb kicks at his stomach, stands with back arched, switches tail, strains to urinate, or dribbles urine, frequently bloody.

Some may recover if the stone is passed soon enough. This blockage of the urinary tract causes pain, colic, and eventually the rupture of the urinary system into the body cavity (hence the name "Water Belly") and death.

If you are watching a lamb that appears to be straining and unable to urinate, put him on a dry floor for a couple of hours. Ordinarily he will urinate in that time, unless there is blockage. Turn the lamb up and feel for a small stone that can be worked gently down the urinary passage. Sometimes manipulation of a small catheter tube (from the drug store) will dislodge the stone.

Veterinarians say that nine times out of ten, the plugging is at the outer end of the urinary passage, so if the stone can be felt right at the end, and cannot be dislodged by gentle pressure, this outer end can be clipped, then disinfected. If the passage can be cleared, and urine spurts out, stop the flow two or three times. It is possible for the bladder to rupture when it is emptied too quickly. If unable to dislodge the stone, a veterinarian may administer a drug such as Testosterone, which has a dilating action, or a smooth muscle relaxer to permit the calculi to pass, or even remove the stone surgically.

Causes:

1. Low water intake. Correct by adding salt to ration, keeping salt and fresh water in the creep. Increasing salt will increase urine volume and decrease incidence of stones.
2. Ration high in phosphorus and potassium—like beet pulp, wheat bran and corn fodder—and low in Vitamin A. Correct by adding ground limestone or dicalcium phosphate, 1 percent or 2 percent of the ration, to make calcium/phosphate ratio approximately 2:1.
3. Crops grown under heavy fertilizer, with high nitrate content, interfere with the carotene in roughage that produces Vitamin A. Vitamin A enrichment of ration would counteract, and some of the lamb creep feeds have this.

4. "Hard" water may be partly the cause. Correct by adding ammonium chloride to feed, approximately ⅕ ounce per head per day, using technical grade. It is a harmless salt, and some pelleted feeds have it in them. Still, calculi are more prevalent when *only* pellets are fed, and seldom develop in lambs who get a level of 20 percent alfalfa.

Rumen papillae on lamb that was on diet of pelleted alfalfa (left) and on diet of chopped hay (right). (USDA Ruminant Nutrition Laboratory)

WHITE MUSCLE DISEASE
("Stiff Lamb")

White muscle disease in lambs is caused by insufficient selenium in the soil, and thus in the feed of the ewe, combined with a deficiency of Vitamin E. When the soil is deficient (as in parts of Montana, Oregon, Michigan, New York, and other areas), then the hay is also deficient in this important trace mineral. Hay from known localities with inadequate selenium should not be fed to ewes after the third month of pregnancy, and during lactation, unless well supplemented by whole grain wheat. Treatment also should include Vitamin E.

To compensate for a known deficiency, or to take precautions just in case, use wheat bran or whole grain wheat as at least 10 percent of the ewe's pre-lambing ration. Wheat in particular can pick up selenium from the soil, and corn also is good. Dried beet pulp would be too, except that it is much too bulky for its nutritive value, so don't give it to ewes in advanced pregnancy who have their stomach space crowded by the growing fetus.

In areas with a *total* absence of selenium, medication can be given to prevent lamb losses. *BO-SE* is an injectable containing both selenium and Vitamin E, and is given to the ewes one to four weeks before lambing, to protect their lambs. *L-SE* is the same kind of medication, given to newborn or one- to two-week-old lambs. These are made in two different strengths. Directions on dosage must be followed very closely. Too much selenium is deadly.

Symptoms of white muscle disease. Lambs have difficulty getting up or walking, as they gradually become affected by muscle paralysis. If treatment is given soon enough, lambs will respond to L-SE.

ENTEROTOXEMIA
("Overeating Disease")

This disease is caused by a multiplying of bacteria called *Clostridium perfringens,* and can strike your biggest and best lambs, those who eat best. It is more common among lambs who are exposed to too much grain and too little roughage (hay), or who have an abrupt change in their feed ration. It also can occur among fairly young lambs who are getting too much milk from a heavy milking ewe. The early creep feeding of both hay and grain will make these lambs not as prone to load up on milk. They may have this disease if grain composes more than 60 percent of their ration, or if they are brought up to a full feed of 1½ or 2 pounds of grain per day too rapidly.

Treatment consists of antibiotics twice a day to lambs, either orally or by injection. See your veterinarian for medication.

Lambs can be given enterotoxemia vaccine ahead of creep feeding time (store all vaccines under refrigeration) in dosage recommended on the bottle, subcutaneously (under the skin) in the woolless area just behind the foreleg.

It also can be prevented by vaccinating the ewes twice before lambing and they will pass on the immunity to their lambs. They are vaccinated with Clostroid DT, which also protects against tetanus, for about three weeks. This is more practical in large flocks. In a small flock, if deaths occur due to enterotoxemia, use of antitoxin (not vaccine) can give lambs temporary protection.

The internal problems in enterotoxemia are basically what happens to mature sheep who accidentally consume too much grain, so keep grain where they cannot get into it.

CHAPTER TWELVE

PROBLEMS OF PREGNANT EWES

PREGNANCY DISEASE,
(Pregnancy Toxemia, Ketosis)

Pregnancy disease is highly fatal if not treated, or if the ewe does not lamb right away. When it occurs, it usually is in the last week or so of pregnancy, and often to twin- or triplet-carrying ewes. It can be reliably diagnosed by urine tests for ketones and acetones, with test strips from the drug store or the veterinarian, but can usually be recognized without this.

Symptoms. Watch for sleepy looking, dopey-acting, dull-eyed ewes, weak in the legs, with a sweet acetonic smell to their breaths. They will probably refuse to eat, then become unable to rise, and will grind their teeth, and breathe rapidly. If treatment is delayed too long, recovery is doubtful.

Treatment. Four ounces of propylene glycol (this is *not* the antifreeze kind of glycol, which is poison), or four ounces of glycerine diluted with warm water, or a commercial preparation for treating ketosis should be given by mouth twice a day. Continue treatment for four days, even though the ewe appears to have recovered, to prevent relapse. Along with the glycol treatment, some veterinarians say that an 8 cc injection of cortisone will shorten the recovery time, but is not necessary.

Keep propylene glycol (or the commercial ketosis medication) on hand before lambing, for prompt treatment of any suspected cases. Subclinical pregnancy toxemia is the same disease, but a mild case where the ewe can be weakened and produce a small lamb or a dead lamb.

Cause. When multiple fetuses demand more nutrients than the ewe can obtain from her food, they are drained from her body. The blood sugar and the sugar in the lymph and other body fluids are taken. Adequate exercise of ewes, *in addition to proper feed,* is necessary to help release glycogen from the skeletal muscle storage in the ewe's body, rather than having her liver depleted, which is damaging. The muscular action of exer-

130

cise also increases the utilization of the ketones that would otherwise poison the ewe (toxemia). Stress is also a cause.

To prevent pregnancy toxemia:

1. Avoid overfatness early in pregnancy.
2. Encourage ewes to exercise.
3. Provide rising level of nutrition in last four to five weeks of pregnancy.
4. Supply a constant source of water.
5. Feed regular amounts at regular times.
6. Give molasses in drinking water.
7. Avoid purchasing ewes too close to lambing.
8. Avoid stress, and hurried driving of pregnant sheep.
9. Make no sudden change in type of grain offered.
10. Give special attention to nutritional needs of old sheep with poor teeth, late in pregnancy.
11. Treat the feet of any lame ewe, or she may not move around well.
12. Give nutritional grain combination, such as ⅓ wheat, ⅓ corn, ⅓ oats.
13. Add molasses to the feed of all ewes, if you have even one case of ketosis.

KETOSIS OR CALCIUM DEFICIENCY?

It can be difficult to tell the difference between pregnancy toxemia and hypocalcemia (milk fever). Pregnancy toxemia can be accurately diagnosed by test strips from the veterinary supply (or ketone sensitive strips from the drug store). But this disease can be a complicating factor in a case of milk fever, so a diagnosis does not rule out calcium deficiency. You can make an intelligent guess by reviewing the circumstances:

If it is before lambing, and there is any possibility that the ewe may not have been fed properly in the last month, it is probably toxemia.

If it is after lambing, and the ewe is providing good milk for twins or triplets, and has had adequate feed with molasses, it is more likely to be primarily milk fever, but could have a trace of pregnancy toxemia as a complication. Most mail order veterinary supplies stock commercial preparations for milk fever which also contain dextrose or other ingredients for ketosis, so they can be used to treat either or both.

MILK FEVER (Lambing Sickness, Hypocalcemia, Calcium Deficiency)

Because so much calcium is needed to form the bones and teeth of the lambs, and so much calcium goes into the ewe's udder of milk, she suddenly may be unable to supply it all, due either to simple deficiency or deficiency triggered by metabolic disturbance, and this deficiency can cause death in a short time. The ewe's lack of sufficient calcium happens more often *after* lambing, but can be just before. It may be brought on by an abrupt change of feed, a period without feed, or a sudden drastic change in the weather.

Symptoms. The onset of the disease is sudden, and progresses rapidly. Earliest signs are excitability, muscle tremors and stilted gait, which are followed by staggering, breathing fast, staring eyes, dullness. The ewe next lies down and is unable to get up, then slips into a coma followed by death. Although this is called a "fever," the temperature is normal or sub-normal, and ears become very cold. To be successful, treatment should start before ewe is down, but even when she is down, prior to coma, there is a chance of recovery.

Treatment. The ewe should be injected subcutaneously with 75–100 cc (divided in five places) of calcium borogluconate or calcium gluconate. The latter is cheaper, easier to find, and equally effective. You can purchase it at a drug store. Tell them it is for livestock use.

The injection could be intravenous, but while sub-cu gives a slower reaction, there is less chance of cardiac arrest (heart failure), and it is a safer procedure at home.

Several commercial veterinary preparations for treatment are sold in mail order catalogs.

If milk fever comes on before lambing, it can be confused with pregnancy toxemia. If it is a calcium shortage, the ewe will show a dramatic improvement after calcium is given.

We had this only once, with a mother of triplets, and her recovery was fast, after treatment.

VAGINAL PROLAPSE

Prolapse of the vagina can occur before lambing, or even following a difficult labor. The vaginal lining will be seen as a red mass, protruding from the genital opening. Do not delay treatment, for it can get progressively worse, and more difficult to repair. Early detection is important, and even though this occurs infrequently, be on the watch for it.

Treatment. If the lining is just barely protruding, confine the ewe in a crate that elevates her hind end, thus decreasing the pressure. Leave her head out to eat, and feed her mostly on grain, plus some green feed (grass, weeds, apples, etc.). Avoid ground up grain or rolled oats. Dust might cause coughing, and aggravate her problem.

A more certain solution, still best begun as early as possible to be successful, is to use a *prolapse loop,* or *prolapse retainer,* which is a flat plastic

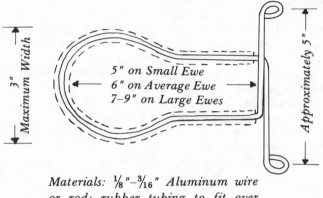

Materials: ⅛"–³⁄₁₆" Aluminum wire
or rod; rubber tubing to fit over
loop.

Homemade prolapse loop. Wire "loop" is covered with soft rubber tubing as a cushion. Put on rubber tubing before making last bends in the wire. Dotted lines show where rubber tubing is placed over wire.

tongue called a *ewe bearing retainer* (available from livestock supply catalogs, see Appendix).

If you have already ordered these ahead of lambing season, you are ready to treat the problem. If not, here is an alternative: Dr. Beck, writing in *Shepherd* magazine gives the following design for a homemade prolapse loop, easily fashioned from wire.

The loop is made from about ¹⁄₁₆ inch wire with only the loop part covered with soft rubber tubing as a cushion, the rubber tubing being put onto the wire before making the final bends. Bend as shown in the drawing. Disinfect the whole loop.

Wash the ewe, place her with her hind end considerably elevated. You can tie the ends of a length of rope to each hind leg, and loop the rope up over the top of a post, or have the rope short enough to just go around your neck, so that the ewe is raised by it. You can hold her steady in this position if she is on her back, with the hind end raised and steadied by a bale of hay.

To replace the vaginal protrusion and insert the loop or retainer:

1. Cinch a rope or belt around her middle so that she cannot strain after you replace the prolapse. A ¼ inch rope around her flank in front of the udder, not so tight that she can't lie down and get up, will do. This will have to be removed when she goes into labor. Sometimes she will stop straining after a couple of days, as the swelling goes down.
2. Wash your hands, and disinfect the loop if you have not done this already.

3. Wash the prolapsed part with cold (not hot) antiseptic water, or put both mild soap and antiseptic in the water to help contract it.
4. Watch out for a flood of urine as you gently replace the vaginal lining. Its bulging may have blocked the opening to the urinary tract. This would cause death if blocked too long.
5. Replace the lining, using lubricant if necessary, and press out all the creases gently. This is much easier with the hind-end-elevated position than it would be if she were lying flat. Even holding her on her back, with her shoulders on the ground and her hind-quarters up against your knee, will relieve much of the pressure against the replacement.
6. Holding the vagina in place with one hand, insert the prolapse retainer or loop, straight in, flat horizontally. If you have made a loop from the pattern given, it should be long enough so that the forward end is in contact with the cervix at the end of the vagina.
7. The loop is held in place by tying it to clumps of wool, or by sutures if the ewe has been sheared or closely crotched.
8. Give a shot of antibiotic to avoid infection.

The ewe can lamb while wearing the loop or retainer, or you can remove it as she goes into labor. It is safer to leave in the loop and try lambing that way, so that the ejection of the lamb does not start her problem all over again.

Mark this ewe for culling, for the prolapse produces permanent damage, and might happen again. Since it may be a genetic weakness, do not keep any of her lambs for breeding.

Sutures. At one time the standard handling of prolapse was with sutures to hold the vagina in. These *must* be removed at lambing time.

We have successfully used dental floss and a curved needle, with pliers to get a good grip on the needle and pull it through.

The safest way to suture is to use only one deep stitch at the top of the vaginal opening, and one across the bottom. Insert the needle from right to left at the top, then bring the needle down and insert it from left to right at the bottom. Knot the two ends together on the right side. The advantage of sewing it this way, rather than crossing the stitches across the center of the opening, is that you can tell when the lamb is coming. There is room for the feet and nose to present themselves, allowing you time to cut the stitching.

And if prolapse occurs after lambing, the replacement and retaining with sutures would be the most logical approach.

I have heard of successful use of a prolapse harness made from a square of auto inner tube, to hold in the prolapse after it is put back into place. I do not recommend it as the most efficient way.

Causes. Prolapse can be caused by any one, or more, of the following:

1. Anatomical weakness, likely inherited.

2. Feeding too much roughage during late pregnancy, with lamb and stomach causing excess pressure, combined with weakness in that area.
3. Dusty hay, making ewe cough.
4. Deficiency of Vitamin A.
5. Extra-fat ewe, lying down in normal sheep position on upward slope, facing upward, causing too much pressure.
6. Pneumonia or lungworms, causing ewe to cough a lot.

ABORTIONS

These can be caused by moldy feed, with mold spores infecting and destroying the placenta, cutting off nourishment to the fetus.

Injury is often a cause, such as when a ram is running with the pregnant ewes, and bumps them away from hay or feed. Narrow doorways, where sheep rush through for feed, are also dangerous as ewes become large with lambs.

Vibriosis is a bacterial infection which can be transmitted during mating, or by ingestion of bacteria from active cases. Abortion usually occurs in the last six weeks of pregnancy, without warning. Aborting ewes should be isolated from the rest of the flock and unaffected ewes moved to a clean pasture, if one is available.

If infection is a problem, a vaccine is available to protect the unaffected ewes.

When a ewe has lost her lamb through abortion in the last few weeks of pregnancy, or has a stillbirth and there is no orphan to graft on her, she should be milked out on the third day and again in a week, if she has a full udder. If the lamb was born dead due to a difficult birth, the first milking can be done at once and the colostrum frozen for future use.

RETAINED AFTERBIRTH
(Retained Placenta)

In almost all cases, the afterbirth comes out normally, usually within the first hour after the lamb is born, depending somewhat on the activity of the ewe. Do not try to pull it out, if it is hanging part way out, as you might cause her to strain and cause prolapse, or to do other injury to herself. You can allow quite some time to pass, without worrying about the afterbirth, for a veterinarian does not consider it a real "retained placenta" until six hours have passed since the birth of the lamb. Some ewes eat the afterbirth if you are not there to remove it, causing you to think it is retained.

If more than six hours have passed, home treatment can consist of an injection of streptomycin or penicillin to ward off infection. Forcible removal of the afterbirth is best done by the veterinarian who can differentiate between the maternal and the fetal cotyledons, to separate them. It is usually not advisable to remove the placenta manually sooner than 48 hours after the birth, and the veterinarian may in the meantime give a drug that could assist in expelling it.

Causes:

1. Exhaustion following difficult lambing.
2. Nutritional disorder, such as deficiency of either magnesium or calcium, which can affect the ability of the uterine muscles to contract properly.
3. Premature birth, sometimes result of poor feeding in the last four weeks of pregnancy.
4. Infection or abortion.
5. Hereditary weakness.

MASTITIS

This is an infection or inflammation of the udder, usually affecting one side, and can be caused by a number of different bacteria.

In severe cases, the ewe has a high fever (105 to 107 degrees) and usually goes off her feed. One side of her udder is hot, swollen, and painful. She will limp, carrying one hind leg as far from the udder as possible. The milk becomes thick and flaky, or full of curds, or watery.

Mild or subclinical mastitis may be undetected, showing up at the ewe's next lambing when she has milk in only half of her udder, and the other half is hard.

Mild cases are most often caused by bruises from large lambs almost weaning age. They bump their mothers with great zest as they nurse, sometimes lifting her hind end right off the ground, or twins who pummel her simultaneously. Mild mastitis has fewer symptoms, and the ewe may just wean the lambs by refusing to let them nurse.

Causes:

1. Undue exposure to cold and rainy weather, lying on cold wet ground.
2. Infection from an active case of mastitis.
3. Udder injury from underbrush or high thresholds.
4. Udder injury from large nursing lambs.
5. Large milking udder, reducing resistance of udder to bruises and infection.
6. Loss of lamb, large milking udder, not milked out occasionally.
7. Sudden weaning of the lambs while ewe still has full milking capacity.

Treatment. One severe type of mastitis will respond if penicillin treatment is given early enough, usually in dosage of 500,000 to 1,000,000 units.

Another type of mastitis results in gangrene. The udder is almost blue, and cold to the touch. Large and repeated doses of Digydrostreptomycin are given.

A veterinarian can examine with a microscope the exudate from the udder and know which organism is the cause, then prescribe proper treatment.

Antibiotics are given by injection, but in some cases antibiotics are inserted into the infected teat. Veterinary supply catalogs sell these, but the only ones we have seen are for cows, and the applicator is a little large for convenient use with sheep. There are combination treatment drugs, for both acute and mild chronic mastitis, and are effective against several of the causative bacteria.

The infected side should be milked out, not onto the ground where it could infect other ewes, but into a container to be poured down the sewer.

The affected half of the udder is not likely to have milk again, even with prompt treatment, so it is best to cull out the ewe.

CHAPTER THIRTEEN

INTERNAL PARASITES

Sheep are very resistant to bacterial and virus diseases, in comparison to other livestock, but vulnerable to internal parasites. One of the principal causes of disease outbreaks is a weakened and run-down condition due to parasite infestation.

Overstocking of pastures leads to severe parasite problems, as well as possible undernourishment of the sheep, which in itself leaves them less resistant to parasites. The highest death losses occur in lambs, yearlings and extremely old sheep, with more deaths when sheep are poorly fed.

If you are starting out with pasture that has never had sheep, or not had them for quite a while, try to avoid the buildup of worms by worming regularly and keeping a phenothiazine-salt mixture available to the sheep at all times.

PARASITE CONTROL

ROTATION OF PASTURES

If you have a few separately fenced pastures, and are able to rotate the sheep from one to another, the worms do not multiply as fast. This must be done in *addition* to giving worming medications.

Rotate pastures every 3½ to 4 weeks if you have four pasture areas. If you have two or three pastures, rotate less often. The eggs of many stomach worms can survive three months in cool or cold weather, but much less in the heat of the summer, so rotation times can be varied.

A freezing winter won't kill off the worm stages on the pasture. Recent studies show it will only preserve them. Therefore, if you have a choice of pastures, don't turn sheep out in the spring on to the last fall-grazed pasture. Select one that was last used in the heat of summer. Unless pasture rotation allows time for the death of the larval stages in the grass, rotation is of more benefit to pasture growth than to worm control.

CLEANLINESS

One step toward better worm control is providing feed and water arrangements that cannot be easily contaminated with manure. The lamb's creep feeder should have a baffle above the feed trough, to prevent the lambs from climbing into their grain.

WORMING

With the development of safer worming drugs such as *Thibenzole* and *Tramisol* and *Loxon,* worming can be done more often and without harming pregnant ewes or small lambs.

Ewes should be wormed two or three weeks before breeding, for if the ewes have a load of parasites, they will not settle properly, and you will have a long lambing period. Also, the ewes will produce fewer twins and more weak lambs, and have less milk for the lambs. Pregnant ewes with worms are drained of needed energy, and their weakness leaves them more susceptible to pneumonia or pregnancy disease, and too weak to withstand a difficult delivery.

One of the peak periods of worms is right after lambing. New Zealand research has concluded that the post-lambing rise in parasites is due to the loss of resistance by the ewe at that time. Worming all ewes two weeks before lambing reduces the hazard to the lambs and avoids the worm explosion after lambing. We worm at least three weeks before lambs are expected. This is not because the worming agent is dangerous, but because the stress of catching and worming in the last two weeks before lambing might combine with some slight nutritional deficiency to trigger pregnancy toxemia in ewes carrying twins or triplets. If this worming is neglected, the worm population in the ewes rises, following the stress of pregnancy and lambing, and leads to a heavy infestation in the lambs.

LAMBS AND PARASITES

Young lambs that are turned out to pasture *with* the ewes, pick up worms that will grow to maturity in about a month. As the worms increase, they cause anemia, and even death. The dying lambs may not be thin, as the effects of severe anemia come on very quickly in the younger animal. You can prevent anemia by worming the lambs at about $2\frac{1}{2}$ to 3 months of age (note withdrawal days before slaughter, on label).

To help prevent severe infestation, practice what is called "forward creep grazing." When rotating pastures, let the lambs graze each clean pasture ahead of the ewes, through a creep-type fence opening of a size that the ewes cannot get through.

We reserve one small pasture of choice grass for the lambs, and locate their Rover Lamb Creep there.

While TM (trace mineralized) salt added to feed does improve the growth of the lambs and the health of the ewes, it also increases the growth of the internal parasites, it has been discovered recently. This can be counteracted by adding dicalcium phosphate to the loose mineralized salt. Since it is such a good source of calcium for the lambs it should be done anyway.

A popular mixture is six pounds TM salt, three pounds dicalcium phosphate, and one pound phenothiazine wormer granules or powder. This low-level feeding of phenothiazine in salt helps control the worm load, between wormings, but is not a substitute for regular worming.

SYMPTOMS

One visible sign of worms is "bottle jaw" (swelling under the jaw). It is a final warning that the sheep have worms severe enough to cause deaths.

Other symptoms are diarrhea (for some kinds of worms) and anemia (for most kinds of worms). Anemia is indicated by the very pale color of the inner lower eyelids and gums, caused by intestinal worms drinking the sheep's blood, as much as a pint a day in heavy infestations.

"Bottle-jaw" from severe stomach worm infestation. (USDA)

There are eight or more kinds of small stomach worms (round worms) that cause anemia but not diarrhea. The sheep become listless, with pale mucous membranes, and lose condition, wasting away and dying if they are not wormed.

The small brownish stomach worm "ostertagia" causes scours. This worm is so perfectly camouflaged against the walls of the sheep's small intestine, that it is difficult to spot in a post mortem.

BROAD-SPECTRUM
WORMERS

Since most infestations are of more than one kind of worm, the broad-spectrum worming drugs are recommended for general worming. If you use two different kinds of wormers, alternate them and use one at one worming and the other at the next, to be most effective.

For most purposes, alternating between two or three of the major broad-spectrum, low-toxicity wormers, Thibenzole, Tramisol, Loxon, and Levasole, will take care of the most prevalent of the stomach parasites.

HOW TO ADMINISTER WORMING MEDICATIONS

Wormer drugs are given as:

1. Boluses — large pills.
2. Drenches — fluid given by mouth.
3. Granules or powders — to be added to salt or to feed.

Boluses. For those without experience, boluses are easier and safer to give than drenches (fluid by mouth). Be sure you are buying the sheep-size bolus, for some wormers are also made in cow size. There are three methods of putting a bolus down a sheep:

1. With one finger, push the bolus until it slips back down the throat. Hold the mouth open far enough so that you don't get your finger crunched by the back teeth.
2. With a bolus gun. (See Chapter 16 on its use, under Oral Medications.)
3. With capsule forceps which hold the bolus (see Appendix). Open the sheep's mouth and deposit the bolus as far back as you can, so she has to swallow it. It is easier to get the bolus far enough back with forceps, as they have a slight angle, so that you can put the bolus over the top of the back of the tongue. After releasing the bolus, withdraw the forceps and hold the sheep's mouth closed. Then wait for a minute to see if she swallows. Don't release the sheep until you know she has swallowed it.

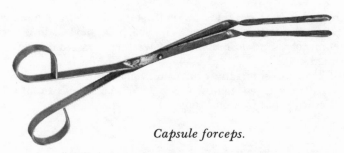

Capsule forceps.

My husband uses the capsule forceps. He straddles the animal, with her neck between his legs, and holds the lower jaw with his left hand, using his thumb as a wedge to hold the mouth open.

Drenches. A ball drench gun, or automatic drench gun, is a liquid wormer, intended primarily for large numbers of animals. The person giving the drench carries a large bottle or cannister of the medication strapped to his back, attached by a tube to the drench gun. Care should be taken not to penetrate the mucous membrane of the throat by overly forceful drenching. Depositing the medicine behind those membranes may cause an abscess. The drench gun is efficient, but an initial expense.

For a small flock, use a handy and inexpensive two-ounce dose syringe, which is a small aluminum pipe attached to a rubber bulb. To administer the liquid drench, have the sheep in a standing position. *Keep its head level,* muzzle not raised above the eyes, or the drench will go into the lungs, causing death. Do not drench while the sheep is coughing or straining. Drenching should be gentle, and slow. For the inexperienced person, drenching is not as safe as worming boluses.

Powders or granules. *Phenothiazine,* usually mixed with salt (ratio one part pheno to ten parts salt) or with trace mineralized salt and dicalcium phosphate, is offered to the sheep "free choice." This low level feeding of phenothiazine is done *in addition* to regular wormings. The pheno-salt box should be protected either in the barn or by a roof shelter out on the pasture. Rain washes the salt and pheno away, and sunshine reduces its potency.

WORMER MEDICATIONS IN COMMON USE

Tramisol®-levamisole, first available in 1971, is effective against three species of stomach worms, six species of intestinal worms, all strains of the barberpole worm, and against lungworm, a type of worm not adequately controlled by other worming drugs. It is safe for pregnant ewes and for young lambs. This medication is sold as oblets, slightly smaller than boluses, in bottles of 24 or 100. Injectable *Tramisol,* which has some advantages in large flocks, has now been approved for sheep use.

Thibenzole® is another of the new and safe wormers, safe even on sheep in a weakened condition. It is effective against most of the stomach and intestinal worms that infest sheep, and actually active against the larval forms of worms, and against the eggs too. A few hours after administering *Thibenzole,* sheep should no longer be passing droppings which will contaminate the pasture. *Thibenzole* for human use is called *Mintezol,* and appears to be the first known cure for trichinosis in humans.

Loxon®, tested for about nine years and released in 1970, is another safe and effective wormer, also active against many of the immature forms as well as the mature parasite. It is more effective than *Thibenzole* against the common stomach worm that causes bottle jaw, anemia and many deaths, and also more effective against Cooper's worm. However, *Loxon* is much less effective against the nodular worm and the large mouth bowel worm. This illustrates the need to alternate wormers, even the broad-spectrum wormers, for a really good control of parasites.

Phenothiazine as a drench or in boluses is no longer a standard treatment. The previously mentioned broad-spectrum wormers are used for seasonal wormings now. *Phenothiazine* is excellent for continuous low-level use in a pheno-salt mix, where it actually renders infertile many of the worm eggs that the ewes drop on the pasture.

LESS COMMON
INTERNAL PARASITES

Lungworms are more prevalent in low-lying or wet pasture, and live in air passages, causing accelerated breathing, coughing, and sometimes a discharge from the nose. The small lungworm (hair lungworm) can cause pneumonia and bronchitis.

Good nutrition helps to build up resistance to the worm (as with other parasites). Keep the sheep away from ponds and wet areas where snails can be found, as several species of slugs and snails act as intermediate hosts for lungworms. When infected sheep cough, eggs are expelled and eaten from the grass by other sheep.

Tramisol given at least once a year will control lungworms.

Tapeworms. The feeding head of the tapeworm injures the intestine, and is thought to facilitate absorption of the toxin involved in enterotoxemia. Tapeworms are not usually the primary worm infestation in a sheep, but since the passed tapeworm segments are large enough to be seen on the sheep droppings, their presence is alarming. Since the manufacture of *Teniazine* and *Teniatol* was discontinued, the elimination of tapeworms has become a problem. The several standard broad-spectrum wormers do not affect them. Wormers containing copper acid arsenate or lead arsenate or copper sulfate are only about 60 percent effective, and quite toxic. A lead-arsenate sheep bolus or drench is officially approved for sheep use.

The *Merck Veterinary Manual* lists *Niclosamide* for tapeworm treatment. This is marketed as *Mansonil*, by Bayer, in some countries, but not here. It is made in this country under the trade name of *Yomesan*, by Bayvet Corporation, and is registered for use against tapeworms in dogs and cats. Bayvet says it has been used experimentally in sheep and shows great promise, but it is not yet registered for that use. Veterinarians who have made experimental use of *Yomesan* say it is extremely safe and effective, dissolves easily in water to use as a drench, but is very expensive, taking about five tablets (total 2500 mgs) for a 100-pound ewe. This is less than the *Merck Manual's* suggested dosage of 35 mg per pound of body weight, but close to the 50 mg per kilogram of body weight that more recent studies conclude to be effective against tapeworms.

There are unconfirmed reports that lambs vaccinated for enterotoxemia seldom have tapeworms.

Nose bots. The nose bot, *Oestrus ovis,* is a fly in its mature form, dark gray and about the size of a bee. The full-grown larvae are thick yellowish-white grubs about one inch long, with dark transverse bands, and found in the frontal sinuses of sheep. When deposited by the fly on the edge of the nostril, the grub is less than $\frac{1}{12}$ of an inch long, and gradually moves up the nasal passages.

During fly season, sheep will put their heads to the ground, stamp and suddenly run with their heads down, to avoid this fly. They become frantic, and press their noses to the ground, or against other sheep as the flies attack them, usually during the heat of the day, letting up in early morning and late in the afternoon. This is more prevalent in areas with a hot summer.

Oestrus ovis *larvae in horn cavity of sheep. (USDA, Livestock Insects Laboratory, Kerrville, TX)*

The head grubs cause irritation as they crawl about in the nostrils and sinuses, and the resulting inflammation causes a thin secretion, becoming quite thick if any infections are present. As the mucous membranes are affected, and the secretions thicken, the sheep has difficulty in breathing, and may sneeze frequently. They can become run down because they lack appetite, or are so annoyed by flies that they cannot graze in peace.

An organophosphate wormer called *Rulex®* was available for treatment of nose bots, but was discontinued. It was effective, but was very toxic when used in improper dosages, and toxic if sheep were kept off the pasture either prior to or after its use.

Merck Veterinary Manual suggests the use of *Ruelene®* as a drench at 45 mg per pound of body weight, which is approximately 2 cc per 10 pounds body weight. While *Rulex®* is no longer available, Dow makes a *Ruelene® Wormer-Drench* for cattle, which is the same formula as *Rulex®*. Its acidity will cause sheep to cough, and it could never be used on a weakened or run-down sheep. Sheep on pasture should be treated and returned promptly to pasture. However, this drench provides essentially complete control of all three developmental stages of nose bots in the proper dosage. While it is extremely toxic at improper dosage, we hope Dow will get this product approved for sheep use, and packaged with exact dosage and cautionary directions for its use, including the advice to treat sheep in the fall, after fly season.

The older treatments, injecting or spraying creosol solutions into the nasal passages, were barbaric and ineffective, and are no longer used. There are medications made abroad, such as *Rafoxanide*. It is sold by Merck Sharpe & Dohme in some countries under the trade name of *Flukanide**, and is said to be effective against both nose bots and liver flukes. It is not now available in this country.

Liver flukes. The three kinds of liver flukes all require an intermediate host, that is, part of the life cycle of the parasite is perpetuated in another creature. In the case of flukes, it is a snail or slug, found on wet marshy land.

Ponds, ditches or swampy land provide the breeding place for the snails, so this kind of pasture is not ideal for sheep. If possible, drain wet areas where snails propagate, or fence off the marshy parts. Snail destroying chemicals can be used, if the area does not drain into waters having fish or used for drinking water for humans or livestock.

As with other forms of parasites, there is sometimes bottle jaw, or pot-belly, during the earlier stages, followed by loss of condition, diarrhea, weakness and death. It can be diagnosed accurately in the liver of a slaughtered sheep, and sometimes can be diagnosed by microscopic examination of feces.

---

*In Australia it is called *Ranide,* and *Ranizole,* which is *Rafoxanide* in combination with *Thiabendazole.*

The most common treatment is carbon tetrachloride, but it may itself damage the liver, and cause death. This damage can be prevented by one of the B vitamins. It has been found that three daily injections, intra-peritoneal, of two grains of vitamin niacin, will adequately protect a sheep for a 50 milliliter dose of the drug, given into the rumen, on the third day. *This is a procedure requiring professional help.*

In addition to *Flukanide*, there is another product called *Hilomid*, also made in Australia and other countries, that is reported to kill mature flukes as well as a large proportion of the immature flukes. It is a concentrated suspension of 3, 4 ', 5 Tribromasalicylanilide, and may be available in this country in the future.

The cost of complying with FDA regulations to obtain official approval of a drug for use in this country may prevent many effective drugs used in other countries from being available here. The cost of getting them approved is more than the manufacturer could hope to profit from their sale, especially if they are for a very limited use such as liver flukes or nose bots.

Coccidiosis. Coccidia are microscopic protozoan parasites, present in most flocks without causing any problem. Outbreaks of coccidiosis are mainly in feedlot lambs of age one to three months, being raised in crowded conditions, but seldom in the pasture arrangement of a farm flock. Any rapid change of feed ration may predispose the lambs to an outbreak of coccidiosis, which usually appears within three weeks of the time they are brought into the feedlot. Other factors are chilling, shipping fatigue, and the interruption of feeding during the shipping time.

Small amounts of coccidial oocysts may be found in most mature sheep, but they seldom show any symptoms of infestation. However, these apparently healthy sheep may be carriers and contaminate their surroundings so that lambs, which are weakened by any change in ration, may be susceptible.

To prevent this, lambs should be fed during shipping, and should not have their ration changed too abruptly from grass to whatever concentrated feed will be given them. Overcrowding and contamination of feed and water must be prevented, for this is the main source of infection.

It once was believed that each species of animals had its own type of coccidia, and there was no cross-infestation. Later experiments have proven that some types of coccidia (there are a number of them even within those that infest sheep) are transmissible to different animal species, that then act as intermediate hosts. Some microscopic cysts in the muscle tissue of cattle or sheep, or even in intestinal tissue, can be fed raw to dogs, with the result that the dogs become infected and pass sporocysts. However, cooking *or* freezing will apparently render these parasites non-infectious, so meat fed to dogs or cats that associate with livestock should be previously cooked or frozen.

Symptoms. The symptoms of coccidiosis are diarrhea, then diarrhea with straining, and finally dark bloody diarrhea, loss of appetite, and some deaths. The lambs which recover are usually considered immune.

Treatment. Treatment usually consists of sulfa drugs, the soluble ones being most effective. Large doses of Vitamin A will hasten recovery.

The Merck Veterinary Manual* gives Prescription #101:

> Nitrofurazone, 4.5 mg per pound body weight, daily for seven to ten days for treatment of coccidiosis; for prevention in exposed lambs, 0.065 percent in feed for twenty-one days.

The Journal of Veterinary Medicine Association reports the experimental use of nitrofurazone at 10 mg per kilogram, daily for seven days, with good results on lambs. It also reports that the administration of two to three grams of sulfamethazine to all lambs daily for one week will reduce the death losses during an outbreak.

This is interesting in light of a January, 1976, report out of the Research Department of the College of Agriculture at the University of Wisconsin. Scientists J.A. Ajayi and A.C. Todd tested a chemical used to control shipping fever in cattle, Aureo-S-700, which is a mixture of sulfamethazine and chlortetracycline. Good results were obtained in particular in field tests, feeding $73\frac{1}{2}$ milligrams daily for six weeks to half of a naturally infected flock of thirty lambs. Seventy days after treatment, the non-medicated lambs had gained an average of 15 pounds, and the treated lambs had gained more than twice as much—34 pounds. Lambs under the same conditions but not infected with coccidia gained around 39 pounds during that period. These studies concluded that Aureo-S-700 would be a good chemical for coccidia control in sheep if it ever becomes feasible for the manufacturer to seek approval of the chemical for that use.

UNAPPROVED DRUGS

There are many medications, especially in the parasite field, that are known to be effective with sheep problems, but are not yet federally "approved" for sheep use.

The question is not just a matter of whether they are safe and effective. The deciding factor is whether it would be *profitable* for the manufacturer to meet the complications and expenses needed to get them registered for use specifically in sheep.

The latest estimate of cost of research and development—involving chemists, microbiologist, toxicologist, physicians, veterinarians and others, and taking as long as $7\frac{1}{2}$ years—averages up to $3.5 million *per product.*

Potential sales volume in the United States does not justify such expenditures, so many good sheep medications are available in Australia and New Zealand but not here. Sheep are *big* business in those countries, and the expense of getting medications approved is warranted there because the cost is supported by the sales volume for the drugs.

**The Merck Veterinary Manual.* Fourth Edition, Copyright©1973 by Merck & Co., Inc., Rahway, New Jersey, USA.

CHAPTER FOURTEEN

EXTERNAL PARASITES

SHEEP TICKS

The sheep tick is not a true tick, but a wingless parasitic fly, known as a *sheep ked,* that passes its whole life cycle on the body of the sheep. It lays little brown pupae which are white inside. These hatch into almost-mature keds in about nineteen days.

Ticks are blood suckers, and roam all over the sheep, puncturing the skin to obtain their food. This causes firm dark nodules to develop and damages the sheepskin, reducing its value. These defects are called "cockle" by leather dealers. Cockle was at one time thought to be possibly a nutritional problem.

The ticks can produce such irritation and itching that sheep rub and scratch and injure their wool, and bite at themselves to relieve their suffering, sometimes becoming habitual wool chewers. They may get impacted rumens from eating the wool. Ticks reduce weight gain by causing varying degrees of anemia, and impair the quality and yield of the meat. The wool value is lower, as the ticks stain the wool with their feces, and the color does not readily scour out. Wool that is tick stained is sometimes referred to as "dingy."

Ticks can be easily eradicated with systematic treatment. The mature tick lays only a single puparia a week, for a total of a dozen or so in her lifetime. The pupa shells are attached to the wool about $\frac{1}{2}$ to 1 inch from the skin. Therefore most of them are removed in shearing, making it easy to eliminate ticks by treating after shearing. The newly hatched ticks die within an hour unless they can suck blood from a sheep. The mature tick cannot survive more than two to four days away from the sheep.

To be effective, all sheep must be treated for ticks at one time, otherwise the untreated ones will pass the ticks back to the treated ones. Examine a new lamb or sheep before turning it in with your own, and treat it if you find even a single tick.

In the nineteenth century, the adult sheep were seldom treated for ticks. Since the shearing was done later in the spring than is common now, the heat of the sun and the scratching of the sheep drove most of the ticks onto the nicely wooled lambs. Herders waited a few weeks after shearing, then

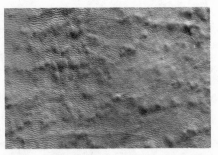

Sheep ticks and what they do: Photo at left shows closeup of keds and puparia including delivery of a single larva. At right, surface of pickled sheepskin showing cockle as firm, dark nodules. (USDA, Agricultural Research Center, Eastern Regional Research Center, Philadelphia)

dipped the lambs in a liquid tobacco dip, sometimes with soap added. The vat used was a narrow box, with a slatted grooved shelf at one side. The lamb was lifted out and laid on the shelf. Then the workmen squeezed the fleece, letting the dip run back into the box. By reusing the dip, five or six pounds of cheap plug tobacco could treat 100 lambs, and was quite effective on the ticks, although the mature sheep still had enough ticks left to get a good start on the next infestation.

Don't make the mistake of leaving any of your sheep with ticks. Every sheep must be treated in one session.

DIP

The standard method in very large herds is to run all the sheep through large dipping vats full of sheep dip liquid, or through spraying vats, where they are given a high-powered spray from several sides at once. This is done usually ten days after shearing, while the wool is still short, but after shearing nicks have healed.

For a small farm flock, this method is hardly practical. It is also unnecessary, for an efficient job of de-ticking requires practically no equipment at all.

SPRAY OR SPRINKLE

Low-pressure sprays, from 100 to 200 pounds per square inch, are ideal for treating sheep when they have been sheared recently and the wool is short.

Sprinkling, with insecticide solution in a garden sprinkler-can, requires very little equipment.

SLOSH

Sloshing wets the animal thoroughly while avoiding many of the disadvantages of spraying and dipping. Do it right after shearing if using rotenone which is not toxic. The sheep do not contaminate the solution with germs that cause infection in a shearing nick, as they are not immersed in the liquid. It is dipped by bucket from a large container of solution and sloshed onto them. We lay them down, roll them on their backs for a good wetting-down of their bellies and undersides and necks. Next, we stand them up and pour the liquid over their backs, giving special attention to the neck region where ticks and pupae are more concentrated. Then we stand back while they shake themselves.

Whatever chemical (report on chemicals to follow) or method you use, if the sheep are badly infested you should repeat the treatment in twenty-three or twenty-four days. If you treat them sooner, you may miss some of the un-hatched pupae, and if you do it later, the unhatched ones have time to hatch and mature and lay their first eggs.

POWDER METHOD

Some of the de-ticking chemicals can be used to powder sheep for tick con-trol, such as Co-Ral on the following list. It is a good method for the follow-up, twenty-four days after sloshing or spraying.

You can thus virtually eliminate all ticks within a short time, with these two treatments, and not have to do it every year, unless you bring in a new sheep, or borrow a ram who has ticks, or loan out your ram to someone whose sheep have ticks. If you loan out your ram, treat him for ticks before returning him to your flock, and you will not have to treat all of them.

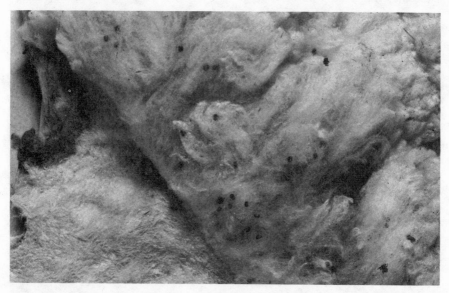

Partially clipped skin showing keds and puparia at the base of the wool. (USDA)

SHEEP KED CONTROL CHEMICALS

Coumaphos (Co-Ral)	1.25 percent in water spray or dip. Or 0.5 percent dust (1 to 2 ounces per sheep). Do not use within 15 days of slaughter. Not for lambs under 3 months old.
Diazinon	0.03 percent in water spray, or 2 percent dust (1½ ounce per sheep). Not within 14 days of slaughter. Not for lambs less than 2 weeks old.
Malathion	0.6 percent water spray or dip (some state regulations restrict this to 0.5 percent water spray or dip), or 5 percent dust (1 to 2 ounces per sheep). Zero days before slaughter. Not to be used on lambs less than 1 month old.
Ronnel (Korlan 24E)	0.5 percent in water spray or dip. Do not use within 28 days of slaughter.
Toxaphene	0.5 percent spray, 0.25 percent dip, or 5 percent dust (1 to 2 ounces per sheep). Do not use within 28 days of slaughter.

These are the regulations on the above chemicals as of mid-1976, but government regulations change from year to year, new chemicals are approved, tolerances change, and old chemicals are sometimes banned. Your agricultural extension agent can usually give you a printed list that is current.

ROTENONE

Until recently *rotenone* was listed in the USDA information pamphlet as "efficient and safe for elimination of sheep ticks with one dipping," but then these external parasiticides became regulated by the Environmental Protection Agency. When EPA requested additional data and studies on the safety and effectiveness of all these products, the manufacturers of *rotenone* evidently did not desire to go to the expense to furnish this information, according to Dr. Eldred E. Kerr. Dr. Kerr, the western regional veterinarian of the FDA, suggested this was due to the competition from many other chemicals. Anyway, by default, *rotenone* is not now approved for official use. It is still an acceptable insecticide for garden vegetables, up to the day you eat them. Being the powdered root of a tropical plant, rather than a manufactured chemical, and an insecticide used by organic gardeners, the sheep raisers who have used it so successfully in the past will probably continue to use it.

For anyone interested in how it was used when still approved for "registered" use:

Rotenone	For dip or slosh, use 8 ounces of the 5 percent wettable powder to 100 gallons of water (about 1½ ounces in a 20-gallon garbage can). Mix it

to a paste in a small amount of water and add it to the large quantity of water in the can. The addition of a small amount of dishwashing detergent will make it more penetrating and effective. Safe for ewes and lambs, up to the time of slaughter. For dusting, use 1.5 percent (garden type) dust, about 2 ounces per sheep.

The effect of *rotenone* is not immediate, so don't be alarmed if ticks are still moving a few hours afterward.

WOOL MAGGOTS
(Fleeceworms)

The maggot problems that are seen in most parts of the country are not the real screwworms that are so destructive in Texas and some other places, but must be contended with in almost all warm summer weather.

Several types of blowflies lay maggot eggs, some are green and iridescent, some have coppery luster bodies, and all are about twice the size of house flies. They appear in the spring, and reproduce from then through hot weather, laying their eggs in masses at the edge of a wound, or in manure-soiled fleeces. The eggs hatch in six to twelve hours, and the larvae feed on the live flesh at the edge of a wound. They enlarge it and can, if not detected, eventually kill the animal.

Maggots need not necessarily be a big deal; the main thing is overcoming one's own revulsion. You can get rid of them quite easily, if they are not too advanced in damaging the sheep. The real danger is not knowing they are there. You can needlessly lose an animal if you do not look at your sheep, catching and examining them if anything looks at all suspicious.

Watch for moist fleece areas, or any injury that may become infested. Notice if animals scratch excessively on fences. This could be maggots or ticks.

When you locate an infestation, clip the wool around it, and spray it with *Screwworm Spray,* or *Wound Guard Smear,* or any of the fly-strike aerosol bombs sold in livestock supply catalogs (see Appendix).

If you don't have one of these sprays, pick out all the maggots you can see and disinfect the wound. One of the sheep tick chemicals can substitute for fly repellent. Even if sprayed with repellent, the sheep should be kept under observation for a few days and treated again if necessary. If the sheep have not been sheared, you might want to shear them after treating the area and removing all the maggots. That would make it easier to spot other infestations. Maggots often infest areas of dog bites, and if your sheep are chased by dogs, check them often for unnoticed wounds and fly strikes.

The wool maggot or fleeceworm (maggot of the blowfly) can be distinguished from the even more dangerous screwworm (maggot of the screwworm fly, *Cochliomyia americana).* You can see the wool maggots move and crawl around, while the screwworms do not move, since they are imbedded in the flesh of the sheep.

Any of the spray medications that will kill screwworms will also be effective on wool maggots.

PREVENT MAGGOTS

The following measures will lessen the chances of trouble with maggots:

1. Keep rear ends of ewes regularly tagged, especially any time that droppings become "loose" due to lush pasture or stomach worms. Worm your sheep regularly. Urine also attracts blow flies, if it soils heavy tags.
2. Treat all cuts or injuries or shearing nicks with fly repellent in hot weather. Injuries or even insect bites invite flies.
3. Put fly repellent on docking and castration sites on lambs in warm weather. Check them periodically until healed. You can avoid the problem, in this instance, by having lambs early in the spring, before hot weather.
4. Use fly traps, such as "Big Sticky," to cut down the number of flies in your barn area.

You can avoid the problem by having lambs early in the spring.

If you have a wool maggot problem with your whole flock, which is unlikely unless they have been attacked by a pack of dogs, you can use one of the sheep dip chemicals on all of them. Ronnel is suggested, a 0.5 percent spray on the sheep, but not within twenty-eight days of slaughter.

COMMON SCAB MITE

Several kinds of parasitic mites produce scab in sheep. The *Psoroptes ovis* is the common scab mite, a pearl gray mite about $\frac{1}{40}$ of an inch long, with four pairs of brownish legs and sharp pointed brownish mouth parts (see illustration, p. 154).

The mites puncture the skin and live on the blood serum that oozes from the punctures. The skin becomes inflamed, then scabby with a gray scaly crust. The wool falls out, leaving large bare areas.

This is not to be confused with the loss of wool that sometimes occurs along the backbone of some breeds of sheep, when kept in areas of heavy rainfall.

To determine whether mites are present, scrape the outer edge of one of the scabs (the mites seek the healthy skin at the edge of the lesions) and put the scrapings on a piece of black paper. In a warm room under bright light, examine the paper with a magnifying glass. The mites become more active when warm, and are visible under the glass.

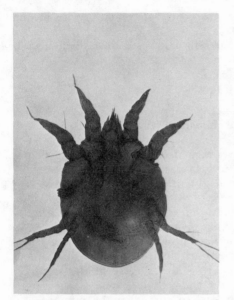 

Psproptes ovis, *the common*   *Sheep ked. (USDA)*
scab mite. (USDA)

The common scab mite, often called *mange mite*, can be eradicated with
a single dipping of 0.06 percent *Lindane*, in states where this drug is per-
mitted. The drug should be used sixty days before slaughter, or follow current restrictions. *Toxaphene* is also effective at 0.05 percent or 0.06 percent
dip, not less than twenty-eight days before slaughter. Since federal regulations and those of individual states change often, you should consult your
veterinarian or agricultural extension agent as to the legal and recommended
chemicals to use. All sheep must be dipped in one session, for the mite is
quite contagious from sheep to sheep.

Infected premises should not be used for clean sheep for thirty days. If
the weather is too cold for dipping, sheep may be treated temporarily
with lukewarm dip, applied with a sponge or brush, until weather permits
a complete treatment.

CHAPTER FIFTEEN

HOOF CARE

If you buy sheep that have hooves in good shape and recently trimmed, they should need trimming only about twice a year.

Many foot disease conditions can be prevented by proper and periodic hoof trimming, most easily done in the spring when hooves are still soft from wet weather, and in the fall after the start of a rainy season. The amount of hoof wear depends on whether soil conditions are mud, sand, or gravel, and whether the barn has a dirt or concrete floor. In some situations, hooves may need trimming more than twice a year, especially when the weather is wet for prolonged periods.

LAMENESS

POSSIBLE CAUSES OF LAMENESS

1. Overgrown untrimmed hooves.
2. Wedge of mud, or stone, or other matter between hoof.
3. Plugged toe gland. Squeeze to remove plug, then disinfect injury.
4. Sprain, strain, nail puncture, thorn.
5. Abnormal foot development. Genetic defect, cull out.
6. Foot abscess.
7. Foot scald.
8. True infectious hoof rot.
9. Vitamin deficiency. Try ADE vitamin in food or injection.

You can help prevent sheep from becoming lame by:

1. Trimming all feet each spring prior to new pasture.
2. Trimming again at shearing time or later in the year. Untrimmed hooves curl under on the sides, and provide pockets for accumulation of moist mud and manure, ideal for growth of foot disease germs.

3. Maintaining dry bedding area during winter.
4. Keeping sheep away from marshy pastures during wet months.
5. Changing location of feeding sites occasionally to prevent accumulation of manure and formation of muddy areas.
6. Having foot bath arrangement for use when needed.

CHECK LIMPING SHEEP

When you notice a sheep limping, try to discover the reason. Notice which foot is being favored, then catch the sheep and trim all four hooves if they need it, doing the sore one last, so as not to spread any possible infection.

Here are the three steps to put a sheep down for hoof trimming: (Upper left) Slip your left thumb into sheep's mouth, back of the incisor teeth, and place the other hand on the sheep's right hip. Bend sheep's head sharply over its right shoulder as you press your hand down and swing sheep toward you. (Upper right) Lower sheep to the ground as you step back. From this position you can lower her to the ground completely, for shearing. (Right) Sheep is raised on rump for hoof trimming.

Photo at left shows hoof edges curling under. Don't let them get worse than this before trimming. At right, they are properly trimmed. The trimming job is easy if you do it twice a year.

HOOF TRIMMING

Using a hoof knife or jackknife, trim the hoof back to the level of the foot pad, so that the sheep can stand firmly and squarely on both claws. The purpose of trimming, other than to prevent lameness, is to give a good flat surface on the bottom, with both pads of the hoof evenly flat. To do this, trim off the excess horn so that it is level with the sole, and also not protruding too far in front. If there are still any pockets where mud and manure can gather, dip these out with the point of your knife or the hook on the end of the hoof knife, and trim the hoof back a little further. Notice the shape of the hooves on your half-grown lambs, for the ideal.

Hoof knives are sold in two sizes, large for cows and smaller for sheep. In dry weather, when feet are drier and harder to trim, hoof shears can be useful for part of the job.

Look for any lump of mud, or a stone or sharp splinter between the claws of the hoof that seems to be sore.

FOOT GLAND

If there is nothing there, check the gland. Sheep have a deep gland between the two toes of each foot, with a small opening at the front of the hoof, on top. These can be readily seen if you look for them. Goats do not have them. The gland's secretion is waxy and has a faint, strange odor, said to scent the grass and reinforce the herding instinct.

If these glands become plugged with mud, the secretion is retained and the foot becomes lame. Squeeze the gland, and sometimes a fairly large blob of waxy substance pops out. If this was the problem, then the sheep should improve.

While on the subject of glands, there are two on the face, just below the inner corner of the eyes. The fatty secretion from these is sometimes noticeable and looks as if it is coming from the eyes. On dark (black sheep) of the Lincoln breed, there is usually a pale gray or white marking at these glands. There are also two mammary pouch glands in the groin, whose scent attracts newborn lambs to the udder.

If there is no evidence of plugged foot gland, or a foreign object between the toes, try to determine if a hoof disease is present. You will have to get a clear idea of what a normal hoof looks like, before you can spot a diseased condition. If you're not familiar with sheep hooves, compare the sore one with another of her feet.

If you are unable to find the cause of the lameness, you can play safe by trimming hooves on all of the sheep. They may need it now anyway. Squirt the hooves with a disinfectant, or one of the foot bath formulas to follow.

We find a foot rot spray medication for cattle, recommended to "fight bacteria" in general, to be very useful for foot irritations. It contains hexachlorphene, dicholorophene, silicone (to hold on the medication) and alcohol. It is available from sheep suppliers. (See Appendix.)

FOOT SCALD

Foot scald is sometimes mistaken for foot rot. In scald, the soft tissue is involved, above and between the hoof. There is inflamed tissue and moistness, and sometimes open sores, usually involving only one foot.

Foot scald is caused by dampness, wet pastures, prolonged walking in mud, or the abrasion of dirt or foreign objects lodged between the toes. The soft tissue between and above the toes and "heel" become irritated and inflamed. It occurs primarily during the wet winter and spring months, and the condition sometimes improves without treatment, in dry weather. It is a major problem only as it lessens foot resistance to more serious disease like abscess or foot rot, and causes lame sheep to eat poorly and not get enough exercise.

Treatment. Trim hooves and spray with antibacterial hoof spray. If no improvement, treat with foot bath—same kind of solution as for hoof rot, or ordinary hydrogen perioxide.

If you do not have footbath facilities, use a large fruit-juice can, filled with 2 inches of the footbath solution, and soak the affected foot for five minutes. Repeat if necessary.

Prevention. Since foot scald is primarily caused by dampness and mud, it helps to get rid of muddy places. See Treatment For Muddy Corrals, at end of this chapter.

FOOT ABSCESS

This is a true abscess, and occurs within the hoof structure, usually afflicting only one foot. It is considered infectious, but not extremely contagious like foot rot.

The infection causes formation of thick pus, and as the internal pressure increases, the sheep becomes more and more lame. Sometimes you can see a swelling above the hoof. Compare it with the other foot. It will be warmer due to the infection.

It is caused by bacteria in manure and dirt, which enter through cuts or a wound, causing an infection of the soft tissues. There is usually a reddening of the tissue between the toes. This infection may become advanced if not treated, and move into the joints and ligaments. At that stage it is almost incurable because it can't be reached.

Abscess is dangerous in pregnant ewes as they will fail to graze, be slow about getting to grain feeding, and not get enough exercise, which can bring on pregnancy toxemia. Insufficient feeding also leads to low birth weight of lambs, and having insufficient milk for them.

Treatment. Unless pressure is released by an incision, the abscess may eventually break and discharge pus. When it is opened or breaks, squeeze out the pus and treat with antiseptic.

New Zealanders clean out the infected area, treat with foot bath, and bandage the foot. They follow this with an injection (intramuscular) of up to a million units of penicillin.

FOOT ROT

Sheep raisers once thought that foot rot was a spontaneous disease of wet weather. It was only about forty years ago that the primary causative bacteria were identified, and found to be capable of surviving only a few days on contaminated pasture.

Foot rot is caused by a combination of two infectious organisms, which enter the foot when irritation, such as foot scald, is present. Because of the combination of germs, it has been difficult to cure by antibiotic injection.

Recently these two causative bacteria have been fully identified and antibiotic treatment, in combination with hoof paring and hoof bath solutions, have made cures more certain.

Symptoms. Foot rot starts with a reddening of the skin between the claws of the hoof. Odor is faint or absent in the beginning, as it is caused by destroyed tissue. The infection starts in the soft horny tissue between the hoof, or on the ball of the heel, then spreads to the inner hoof wall. By this time it has developed a strong unpleasant odor. As the disease progresses, the surface of the tissue between the underrun horn has a slimy appearance. The horny tissue of the claws becomes partly detached, and the separation of the hoof wall from the underlying tissue lets the claw become misshapen and deformed. There is relatively little soft tissue swelling. In severe infections, it is often more practical to dispose of the most seriously affected animals and concentrate treatment on the milder cases.

Treatment:

1. Hoof trimming, trimming off as much of the affected part as possible, and exposing infected areas to the foot bath. Germs grow in the absence of oxygen, so this paring is important. Disinfect knife after each hoof trim.
2. Run sheep through foot bath, with formalin solution, or copper sulfate solution, formulas to follow.
3. Combiotic injection, listed in Foot Rot Drug Treatment.

The combined hoof trimming, foot bath and injection are said to cure 90 percent of the cases, *if* sheep are held on a dry yard or pasture for *twenty-four hours.*

TREATMENTS

FOOT BATH

If you run the sheep through a trough of plain water first, it keeps the bacterial bath clean longer. Be sure sheep have had water and are not thirsty, so they do not drink any of the foot bath.

Feet should be trimmed before the foot bath, as both of these preparations will harden and dry out the hooves, making them very difficult to trim. Dip your hoof knife in disinfectant between each hoof and each sheep, so you do not needlessly spread germs. Burn the hoof trimmings.

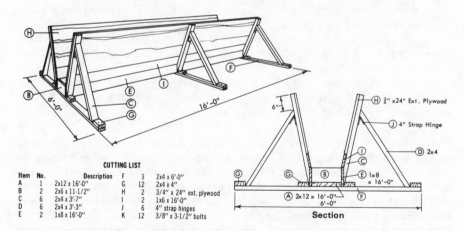

CUTTING LIST

Item	No.	Description			
A	1	2x12 x 16'-0"	F	3	2x4 x 6'-0"
B	2	2x6 x 11-1/2"	G	12	2x4 x 4"
C	6	2x4 x 3'-7"	H	2	3/4" x 24" ext. plywood
D	6	2x4 x 3'-3"	I	2	1x6 x 16'-0"
E	2	1x8 x 16'-0"	J	6	4" strap hinges
			K	12	3/8" x 3-1/2" bolts

Foot bath trough is useful if hoof problems develop, and the foot bath is combined with proper trimming. This trough need not have sides made of plywood. Ship lap or 1 x 6s can be used. A gate is needed at the exit to keep sheep standing in the bath for the required time. (Midwest Plan Service)

Trim non-limpers first, and run them through the foot bath first, then turn them into clean pasture (not grazed for at least two weeks) if you have one. Next run the limpers through, and keep them in a dry area if possible, treating them regularly every three or four days, or have them walk through the bath on the way to daily feeding.

FOOT BATH FORMULAS

Copper sulfate. Use 25 pounds of copper sulfate to 10 gallons of water. The copper sulfate dissolves better in warm water. Do not store this solution in metal containers, as it is corrosive. The solution in the trough should be about two inches deep.

Use as a walk-through bath, several times a week. Severely limping animals can be held in it for five minutes. The solution is toxic to nursing lambs, if splashed on the ewes' udders, so do not rush the sheep, making them splash it. It also stains the wool.

Formaldehyde (Formalin). Use one gallon standard strength (37 to 40 percent) formaldehyde to 12 gallons of water. Keep it about two inches deep in the foot bath. This is called a 5–10 percent bath. It has a very offensive odor, but is non-corrosive in containers, easy to prepare and does not stain the wool. Add a little diesel fuel to the foot bath to reduce the toxicity of the fumes from the bath, if they bother you.

Both copper sulfate and formaldehyde are toxic to nursing lambs. The formalin bath should not be used more than once a week.

Paraformaldehyde. *Shepherd* magazine mentions a formaldehyde derivative, paraformaldehyde, carried by chemical suppliers as a bathroom disinfectant, which can be purchased in flake form. It is sprinkled on the ground around waterers on a regular basis about once a week, and has been used in other countries to halt the spread of hoof rot. It is still necessary to pare the hooves, and still desirable to foot-bathe them.

ALTERNATIVE TO FOOT BATH TROUGH

If you have only three or four sheep, it may not seem practical to build the foot bath arrangement shown in this chapter. You can use a large empty fruit-juice or coffee can, with foot bath mixture to a depth of two inches. For each sheep, use it on the lame foot, and as a precaution also on trimmed healthy feet, if you feel the infection is spreading. Apply the foot bath with a brush to the hoof, then hold the infected foot in the can of solution. If you are treating a front foot, hold up the other front foot, forcing the animal to stand on the foot that is in the bath. Keep the animal a minimum of a half-hour on a dry floor before turning out to pasture, preferably a dry pasture. Repeat in a week, if still limping.

FOOT ROT DRUG TREATMENT

After hoof trimming and foot bath, the injection of combiotic can be given. Suggested dosage has been 2 million units of penicillin and dihydro-streptomycin, intramuscularly in the leg (not within thirty days of slaughter). This injection, plus foot trim and foot bath, has been proven most effective.

There are some commercial preparations cleared for use with cattle, not yet released by the FDA for use with sheep, that may be helpful in the future. One is an iodine product containing ethylene diamine di-iodide.* Although foot rot in cattle is not caused by the same organism as in sheep, it does respond to some of the same treatments.

VACCINE

A vaccine has been perfected and is in use in Britain, New Zealand, and Australia, and in clinical trial tests in Canada. It was hoped that this vaccine would be available in this country, but there is no conclusive proof that it is effective on the strains of *Bacteroides nodosus* (the primary of the two causative bacteria) present in the United States.

Oregon State University has been conducting tests for several years in an attempt to adapt the vaccine to the forty-two strains of *B. nodosus* detected in Oregon flocks. Only five of the bacteria isolated at OSU belong to any of the major types identified in Australia.

Of the eight flocks involved in the 1974–75 Oregon field tests, the Australian vaccine reduced the incidence and severity of hoof rot in only three flocks. Research is continuing at OSU, and elsewhere, to develop a vaccine that is effective under conditions in this country.

TREATMENT FOR MUDDY CORRAL

Nothing is harder on the feet of sheep than the deep muddy yards that develop around barns in wet climates. It is possible to "dry up" muddy corrals which are such a problem in wet weather.

The Department of Primary Industry in Queensland, Australia, says the use of slaked lime will convert a wet muddy yard to a solid surface in about an hour. It must be done while the area is wet, and takes about a 50-pound bag of slaked lime to each 3¾ quarter yards. Grass should be removed, and soil broken up to a depth of six inches. In our wet season (fall through spring) there is no grass on these muddy places, and it is just plain mud for a depth of six inches.

Lime, when spread, should be mixed in as soon as possible, for its action on the soil starts at once. A rotary hoe can be used for mixing. After this, the area should be compacted, and a farm tractor is most suitable. As the tractor is driven around the area, the soil will become visibly firmer, and compaction should continue until the surface is quite firm.

---

*Listed as Diamine-Iodide in *Kansas Vaccine Catalog*.

CHAPTER SIXTEEN

MEDICATION

Successful treatment of any sheep illness requires detection as early as possible, before the sheep is "down." With the discovery of new medications and antibiotics, it is no longer true that "a down sheep is a dead sheep," but the chance for recovery is much better if illness is diagnosed and treated before it has progressed very far. Some of the causes of illness are: lack of exercise, unsanitary housing, moldy or spoiled feed, toxic plants or other substances, improper diet (insufficient feed or overeating), parasites, injuries, infection from assisted lambing, germs from other sick sheep, and stress (weather, shipping, predators, etc.).

HOW TO DETECT
A SICK SHEEP

You have to be familiar with the normal behavior of your sheep, even for each particular animal, to know when one is acting abnormally. Have some quick and easy way of catching them when needed, like a corral where they can be fed and then enclosed.

Signs of abnormality are loss of appetite, or not coming to eat as usual. Be concerned if a sheep is lying down most of the time, when the others are not. Any weakness or staggering, or unusually labored or fast breathing, change in bowel movements, or temperature over 104° is indication of possible illness.

Sheep temperature is normally in the range of from 100.9° to 103°, and the average normal (rectal) temperature is 102.3°. A veterinary rectal thermometer has a ring or a hole at the outer end, where you can tie a string.

If you have to catch a urine sample, such as for use with the pregnancy ketosis test strips, try a plastic cup fastened to the end of a shepherd's crook handle. Impatient? Try holding the sheep's nostrils for a moment. This sometimes triggers urination.

AREAS OF GERM TRANSMISSION

1. Watering places, with water contaminated by sheep droppings, or feces from another species. Respiratory disease may also be spread by nasal discharge into drinking water.
2. Uncrotched wool on a ewe with a lamb.
3. Manure accumulated in a lambing shed, or around feeding trough.
4. Feeding areas on the ground, except through fence lines where food cannot be trampled by the sheep.
5. Wet muddy places, which encourage and transmit hoof ailments.
6. Newly acquired sheep.
7. Venereal exposure at breeding time.
8. Insect, bird, snail, dog, and other hosts of parasites, and carriers.

Maintain sanitary surroundings. If you keep sheep adequately fed, they are less apt to eat poisonous substances, as well as being better able to withstand disease. If you worm them regularly (see chapter on internal parasites) there is less of a parasite build-up to weaken them and leave them susceptible to disease.

DRUGS

It is a good practice to have a supply of standard medicines on hand for use in an emergency. These include bloat medication, antibiotics, propylene glycol for pregnancy toxemia, Cal-phos or other treatment for milk fever, iodine and disinfectants, mineral oil for constipation, dextrose solutions and hoof footbath preparation. Of the antibiotics, penicillin and sulfa will halt many infections. For certain infections, other antibiotics are prescribed.

Penicillin is quite safe to use, for its toxicity in sheep is extremely low. The severe allergic reactions that sometimes occur in humans are almost unknown in sheep. It is of use in pneumonia, infection after lambing, as a preventative against infection in maggot infestation (after cleaning and dressing), and is some help for enterotoxemia.

The sulfa drugs are also useful, but an ample supply of water must be available for a sheep being given any sulfanamide. A toxic side effect is possible with sulfa.

In an extreme emergency situation, when you must use a medication meant for cows, it is generally estimated that the dose for one cow would be a dose for about five sheep. Since in many drugs the exact dosage is very important, with even a slight overdose being fatal, it is best to use sheep medicines, or have the advice of your veterinarian. He can prescribe medicines not "registered" for sheep use, if he wants to take the responsibility.

METHODS OF ADMINISTERING
DRUGS AND MEDICINES

1. *Oral,* by mouth such as worm boluses with bolus gun, or with capsule forceps.
2. *Oral,* powder such as vitamins, placed well back on the tongue for treatment of an individual animal, or in feed or drinking water for general treatment of whole flock.
3. *Oral,* liquid given as a drench with syringe, or in drinking water.
4. *Spray on,* such as pink-eye spray, or maggot or screwworm bomb.
5. *Pour on,* such as iodine on newborn lamb navel, or disinfectant on minor wounds.
6. *Subcutaneous,* medication injected just under the skin.
7. *Intradermal,* medication injected into the skin.
8. *Intramuscular,* liquid injected into heavy muscle, as antibiotics.
9. *Pessaries,* as uterine boluses to prevent infection after an assisted lambing.
10. *Intramammary,* injection of fluid or ointment into teat opening, as mastitis drugs.
11. *Intraperitoneal,* injection of liquid through *right* flank into body cavity.
12. *Intraruminal,* injection of fluid into rumen, on the *left* side, as for bloat remedy when too late to be given by mouth.
13. *Intravenous,* injection of fluid in vein.

The last mode of medication obviously should be done by a veterinarian or very experienced person. Numbers 11 and 12 are surgical procedures, and should be done by an experienced person or certainly with the help of someone who has done them before.

ORAL MEDICATIONS

Boluses (large pills) will slip down easier if coated with oil. Do not soak them or they will disintegrate. Just coat lightly before giving them. (See Chapter 13 on giving them.)

The easiest way to hold the sheep is to back it into a corner and straddle it, facing forward. You can hold the bolus in a "bolus gun" and eject it when you have the pill in the right location back beyond the base of the tongue. Or it may be easier to deposit the bolus far back into the throat with "capsule forceps," which are like a pair of pliers with long angled handles. You will have to wedge the mouth open with the left thumb while inserting the bolus gun or forceps with the other hand. Don't release the sheep until you are sure the medication has been swallowed.

Liquid medication is given with a dose syringe, or a dose gun for larger numbers of sheep. The pipe or nozzle on your dose syringe should be about five to six inches long, and have a smoothly rounded tip that will not injure the sheep. The sheep's head should be in a level position, with the nose no higher than the eyes, so that the liquid will not go into the lungs and cause pneumonia. The safest procedure is the "trickle" method. Administer the liquid slowly, holding the sheep's head in the level position.

INJECTIONS, GENERAL INFORMATION

Sterile procedures must be maintained, to avoid serious infections. Use clean and sterile syringes and needles. The best method of sterilization is by boiling twenty minutes. Store hypodermic needles in alcohol between uses.

Disposable plastic syringes can be ordered from veterinary supply catalogs (see Sources chapter) or in some states can be obtained from your local drug store, and are inexpensive.

To fill a syringe with medication, swirl the bottle to mix it without causing bubbles. Before withdrawing the medication, sterilize the rubber stopper of the bottle. Then sterilize the syringe and needle, and fill the syringe while holding the bottle upside down, so you do not draw air into the syringe. To be doubly safe you can replace that needle with another sterile needle before making the injection. Do not let the needle touch anything, or it will no longer be sterile. If possible, have a helper to hold the sheep or hand you the necessary medicine and equipment.

Protect drugs from freezing and from heat. Many medicines require temperatures above freezing, and below 50 degrees. Read the label on each medication for directions as to the ideal storage temperature. Many antibiotics require refrigeration. Check the expiration date on the package.

Read the dosage carefully and follow it, or use according to the advice of your veterinarian. On some drugs, there is not much leeway between the dose required in order to be effective and the overdose that would be fatal or harmful.

Subcutaneous injection. Subcutaneous means depositing medication directly beneath the skin. The medication usually should be at body temperature, especially with young lambs, and can be given in the neck, but the preferred place is in the loose, hairless skin below the armpits, which can be better disinfected. A dosage of more than 10 cc is best distributed among several sites instead of all in one place (use less with lambs).

Swab the area with alcohol or other antiseptic before and after an injection. Rub the area afterward, to distribute the medication and hasten its absorption.

To inject, pinch up a fold of loose skin of the armpit. Insert the needle into the space under the skin, holding the needle parallel to the body surface.

Do not make the injection near a joint, or in areas carrying more than a minimal amount of fat under the skin. With this injection there is a little problem with veins, but if you want to make sure you are not in a vein, the plunger can be pulled out very slightly before making the injection. If it draws out blood, try another spot. Medication for sub-cu use should not be injected into a muscle.

The correct method of injection is to pull away a fold of loose skin of the neck, insert needle into space under the skin, and avoid injecting into the muscle, near a joint or in areas carrying more than a minimal amount of fat under the skin. The fold of skin is called the sub-cu "tent."

Intradermal (sometimes called intracutaneous). An intradermal injection is into the skin instead of beneath it like sub-cu, and is rarely used. The inserted needle is so close to the surface that it can be seen through the outer layer of skin. Sites used are the same as for subcutaneous.

Use a very fine needle, 20 gauge, and about ¾ inch long. Pinch up the skin, hold the needle parallel to the skin, and insert it into the skin layers. Inject slowly while drawing out the needle, which will distribute the dose along the needle's course.

In the case of a vaccination, the fluid will raise a lump, which will become enlarged in a few days. (Actually, most vaccines are applied by scratching the skin, by sub-cu, or by intramuscular injection.) Follow the directions of the manufacturer or veterinarian very carefully both as to the dosage and the manner of administering. Vaccination sites on lambs would ordinarily be the side of the breast bone, and on ewes the underside of the tail or the non-wooled area behind the elbow (medial side of thigh). However, exact directions must be followed, for best results.

Intramuscular injection. Fresh antibiotics are important, as are a sterile needle and sterile procedure to avoid risk of deep-seated infections. Keep the needles in rubbing alcohol when not in use, and boil the syringe before using. Twenty minutes of boiling seems like a long time, but it takes that long to kill the most stubborn germs.

Check old needles for sharpness. Swab the injection site with alcohol, and the top of the bottle stopper before withdrawing the medication through the stopper.

An intramuscular injection deposits the medication deep into a large muscle, such as in the neck or heavy muscle of the thigh. If you can get an experienced person to demonstrate this, you can see the exact site that will avoid both a nerve and the best cuts of meat.

With an assistant holding the sheep still, thrust the needle quickly into the muscle. To be sure it is not in a vein or blood vessel, pull the plunger out just a bit. If blood is sucked into the syringe, try another site.

Give correct dosage of fresh medication, and with a very sharp needle to avoid tissue damage. It is usually best not to inject more than 10 cc into one spot, in a sheep.

Intramammary. For udder ailments such as mastitis, injections of liquids or ointments are sometimes administered into the teat. The nozzle or tube of udder antibiotics is designed for cows, and is not easy to use on sheep.

Wash your hands and the udder, then carefully disinfect the teats, or germs will be carried in when you use the medication.

Remove the cap of the infusion tube, and insert it into the teat canal gently. Don't remove the cap until you are completely ready to use it, for you want to avoid bacterial contamination, which could complicate an already serious condition. Squeeze the dose into the teat, then massage the dose upward toward the udder.

Intraperitoneal injection. This should be done only by the person familiar with aseptic technique or anatomy. Complications are common after this procedure. If it must be done in an extreme emergency, have an experienced person guide you. It is easier if one person holds the sheep, straddling it just in front of the shoulders.

Medication should be at sheep body temperature, with sterile syringe of 25 cc or 50 cc, and sterile 16 gauge needle, 1½ inches long. Disinfect the bottle stopper before withdrawing medication, and use another sterile needle to give the medication, to avoid the possibility of introducing any infection into the body cavity.

Clip the wool from the right flank, in the shallow triangular depression below the spine, between the last rib and the point of the hip bone. Medication injected into the center of this depression goes into the peritoneal cavity, a sac-like cavity in the abdomen.

Scrub the injection area with soap, rinse, dry, and then disinfect the skin with alcohol or iodine. Hold the needle perpendicular to the skin, pointed toward the center of the body. Inject quickly the full length of the needle, and eject the medication. If it does not eject easily, the needle may be clogged with a plug of tissue, or may not be in the right place. If so, withdraw the needle, replace it with a newly disinfected needle, and try again. Rub the injection site with alcohol or some other disinfectant afterwards.

ANTI-SERUMS

Anti-serum* (often called "serum") is from the blood of hyperimmune animals (animals that have a high specific antibody protective level against a certain disease).

When the anti-serum is injected, a temporary or passive immunity results for a variable period—usually from ten to twenty-one days.

---

*Definitions of anti-serums, bacterins, vaccines, and antibiotics courtesy of Kansas City Vaccine Co.

Anti-serum is used to protect uninfected animals for a short time when disease is present in the herd and to treat infected animals as an aid in overcoming disease. It is given to healthy animals along with vaccine to give temporary protection while the vaccine causes a permanent immunity to develop.

BACTERINS

Bacterins are suspensions of bacteria which have been grown in culture media, and chemically or heat killed. They are not capable of producing disease and can be used without danger of spreading disease. The bacteria used in the production of the various bacterins are of strains isolated from animals having the diseases against which protection is desired.

Bacterins are suggested as an aid in establishing immunity to specific diseases. One dose is usually employed for this purpose on uninfected animals; however, a second dose administered four to five days later will tend to make the immunity much stronger. In infected herds, two or three doses of increasing size are commonly given at three- to four-day intervals as an aid in controlling the disease. Maximum protection is usually for up to a year.

VACCINES

Vaccine is a biological preparation which, when injected into the animal, causes the animal's body to build its own protective antibodies against a specific disease. Some vaccines give immediate and long-lasting protection. Some bacterins cause similar protective antibodies.

ANTIBIOTICS

Antibiotics is the general term for a group of products which interfere with bacterial growth. This explains why certain diseases of bacterial origin can be treated by antibiotics. Antibiotics, when given in adequate concentration, can kill some forms of bacteria. Low concentration (below recommended levels) may only inhibit the growing bacteria and not aid in correcting the disease.

TOXIC SUBSTANCES THAT
CAN CAUSE ILLNESS OR DEATH

1. Waste motor oil, disposed of carelessly.
2. Old crankcase oil (high lead content).
3. Old radiator coolant.
4. Orchard spray dripped onto the grass.
5. Weed spray (some have salty taste).

6. Most sheep dips.
7. Old pesticide or herbicide containers, filled with rainwater.
8. "Empty" lead buckets, filling with rain water. (Rain water is soft, readily dissolves enough lead to kill a sheep that drinks from it.)
9. Salt: Sheep require common salt for good health, and it should always be available to them. If they are deprived of it, then are allowed free access to it again, they may ingest large quantities. Salt poisoning symptoms are trembling and leg weakness, nervous symptoms and great thirst. Treatment: Lots of clean, fresh water to drink.
10. Commercial fertilizer: It has high nitrate content, and in the rumen of sheep is converted to nitrite, causing death. Be careful not to spill any fertilizer where sheep can eat it, and store the bags away from sheep. They may nibble on empty bags. Several rainfalls are needed after fertilizing a field, and it still may not be safe unless the pasture is supplemented with grain and hay. Symptoms are weakness, rapid open-mouth breathing, and convulsions. For a home remedy, use vinegar, one cup per ewe as a drench. A veterinarian will have a more certain treatment, if started soon enough.

CHAPTER SEVENTEEN

WOOL AND SHEARING

SHEARING

Many county agricultural services sponsor shearing schools in early spring for one or two days at a nominal fee. They usually limit their instruction to electric shearing, but what you would learn there would be valuable with either electric or hand shears. We have used both, and prefer the hand shears both for our own sheep and those we shear around the neighborhood.

"Rigged" shears have a leather strap taped onto the left handle (for right handed use) and a rubber stop is taped to the top of the right handle, at the base of the blade. These shears are more comfortable to use, and the strap prevents them from being kicked out of your hand.

ADVANTAGES OF HAND SHEARS (BLADES)

1. Inexpensive way to get started. Order from any sheep mail order catalog.
2. Need no electricity. Shear any place.
3. Easy and quick to sharpen, with a hand stone.
4. Very light weight, to carry with you.
5. Do not shave the sheep close, so no risk of death from loss of body heat in cold, driving rain or sudden storm.

SHARPENING HAND SHEARS

To sharpen, reverse the normal position of the blades, crossing over each other. Using a medium sharpening stone, follow the *existing* bevel of each blade, with long strokes. Do not sharpen the "inside" surface of either blade. If there are any slightly rough edges when you are through sharpening, run the stone flatways along the inside surface of the back (not the edge) to remove the edge burrs. For touch-up sharpening while shearing, close the shears firmly so that each cutting edge protrudes beyond the back of the other blade. Using the fine side of a small ax-stone, follow the existing bevel of each blade.

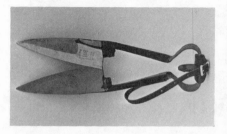

"Rigged" sheep shears.

Sharpening hand shears.

HOW TO SHEAR

The "trick" in shearing is not just the pattern of the shearing strokes, which lessens the time involved in removing the wool, but is the immobilizing of the sheep by the various "holds" that give the sheep no leverage to struggle. A helpless sheep is a very quiet sheep. This cannot be done by the use of force alone, for forcible holding will make the sheep struggle more. Try to stay relaxed while you work.

Note both the holds on the sheep, often by use of the shearer's foot or knee, and also the pattern of shearing in these photos.

You may like to use a *shearing belt* to lessen the strain on your back. This one is cut from a discarded inner tube and has a wide buckle. The width of the available buckle determines the width to cut the belt at both ends, and it should be wider in the middle part which fits across your back.

At one time, shearing cuts were dabbed with tar, to help them heal and to keep away flies. Shearing cuts heal quickly, but use an antibacterial spray, for they can become infected, sometimes resulting in the infection spreading to the lymph glands. While commercial shearers do not do this, a person shearing his own sheep will have more incentive to do it.

This shearing belt can be cut from an inner tube.

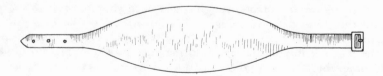

Slip left thumb into sheep's mouth, back of the incisor teeth, and place other hand on sheep's right hip.

Bend sheep's head sharply over her left shoulder, and swing sheep toward you.

Lower sheep to the ground as you step back. From this position you can lower her flat on the ground, or set her up on her rump for foot trimming.

Start by shearing brisket, and up into right shoulder area. One knee behind sheep's back, other foot in front.

173

Sheep is on her left side. Trim top of head, then hold one ear, and shear down cheek and side of the neck as far as the shoulder, into the opening you made at the brisket.

Place sheep on her rump, resting against your legs. Shear down the shoulder while she is in this position.

With sheep in this position, and with you holding her head as shown, shear down the left side.

Hold her left front leg up toward her neck, and from this position shear her side and belly.

With only a minor shift in the position of the sheep, you are now ready to shear the back flank.

By pressing down on the back flank, the leg will be straightened out, making it easier to shear.

From this position, the sheep is shorne along her backbone, and a few inches beyond, if possible.

By holding up the left leg it is possible to trim the area around the crotch.

The job is half done. The shearer's feet are so close to the sheep's belly that she cannot get up.

Holding one ear, you start down the right side of the neck. No, you don't hold the ear tightly enough to hurt her.

Shearer holds sheep with left hand under her chin and around her neck, and shears the right shoulder.

Sheep is pulled up against the shearer, to expose her right side, so that he can shear down that side.

Shifting position, as shown in this photo, he shears further down the side and the rump.

Shifting his position, he finishes the right flank and shears the sheep's rear end.

He again moves his position, and, holding up the rear leg, he shears the right side of the crotch.

The job is done, the sheep is back on her feet, and within a minute, is eating grass again.

SHEARING SUGGESTIONS

1. Shear as early as the weather permits. Shearing nicks will heal before fly season. Ewes can be sheared before lambing, making it easier to help the ewe if necessary, and removing dirty wool tags the lamb might suck on.
2. Never shear when the wool is wet or damp. It is very hard to dry it enough to sack or store. Damp wool is combustible, and can even mildew.
3. Shear on a clean tarp, shaken out after each sheep, or on a floor that can be swept off.
4. Shear fleece in one piece. Don't trim legs or hooves onto the fleece.
5. Avoid making "second cuts," going twice over the same place to tidy up, or overlapping your strokes.
6. Remove dung tags, and do not tie in with the fleece.
7. Roll fleece properly, and tie with paper twine, if selling to a wool dealer or in wool pool.
8. If selling to handspinner, pack unrolled fleece gently into empty paper feed bag, one or two fleeces to a bag. You can shake out much of the junk and second cuts before bagging to make fleece more valuable.
9. For spinning wool, get top price for top quality (clean fleeces without manure tags and vegetation).
10. Ask lower price for lower quality fleeces, and explain the lower price to the customers.

THE WOOL

ROLL AND TIE FLEECE

The acceptable method of rolling, in this country, is to spread out the fleece, skin side down, and fold side edges in toward the middle. Then, fold neck edge in toward the center. Last, start rolling from tail-end of fleece, and make a compact roll. Using paper twine, tie around one direction, cross the twine and tie around the other direction, and knot securely. A slip knot in the starting end of the twine makes it easy to cinch it tight. Then loop the twine around the other direction and knot it.

WOOL FOR HANDSPINNERS

If you set aside your best fleeces to sell for handspinning, be sure they are absolutely dry. They can be stored for a while in a plastic bag, but this is not best for long-term storage. With nice fleeces to sell, it seems unlikely that you will have to hold them very long before they are sold. Find where the nearest craft classes are given, and let it be known that you have fleeces to sell.

In addition to being free from seeds, burrs, stained wool and tags, the fleece should not have "weak staple," from illness or other cause, as this would make it worthless for spinning.

Weak staple can be caused by illness, pregnancy, undernourishment, or poor nutrition, and is called "tender" wool. To test for weakness, stretch a small tuft of wool between both hands. Strum it with the index finger of one hand. A sound staple will give a faint, dull, twanging sound, and will not tear or break.

SHEEP COATS

One way of having clean fleece is to put a coat (or sheep blanket, as they are called in supply catalogs) on your sheep. Sheep coats have been tested in Australia and New Zealand, and in this country mainly at the University of Wyoming.

The protection of the coat makes increased production of *clean* wool, which is expected, but it also results in from 13 to 27 percent longer staple length of wool, reflects improved body weight under harsh range conditions, and makes shearing much easier, partly because the fleece is cleaner. In areas of severe winters, the sheep can conserve energy. This shows up in the maximum percentage of increased wool growth. Mild winters would probably reduce this gain.

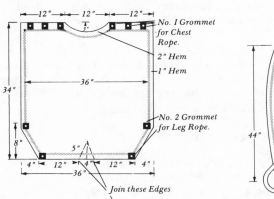
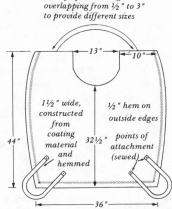

These are two styles of coats to protect sheep, with Number 10 duck or canvas used in most cases. The right pattern can be made in three sizes, with the large having a half-inch overlap on the neck flap, and 27-inch leg loops. Medium has a 1½-inch overlap and 24-inch loops, while the small has a 3-inch overlap and 24-inch loops. The loops are 1½-inch strips of the coat material, hemmed. Left pattern has grommets used for the chest and leg ropes.

One sheepman reported no death losses from coyotes during coat use, and thought perhaps the sound created by the plastic coats as sheep moved, or the sight of the different-colored coats, warded off the predators.

Cost seemed the main factor in making sheep coats less than practical. Cotton coats were not durable around barbed wire or brushy pasture. Sturdy nylon-based coats were more durable, but had the disadvantage of making the sheep sweat during warm weather or close confinement.

The coats were of greatest help in protecting the fleece under conditions that were also very hard on the coats, many of them not lasting out the five or six months of the tests. They were considered too expensive if they could not be made to last for several seasons of wear. In small farm flocks, which do not encounter the hardships of range sheep, they should last a lot longer. With catalog prices of around $10 for ones made out of eight-ounce duck, you might want to try your hand at making one of these designs.

BEAUTY OF FLEECE

Heredity determines the wool type, but its quality and strength depend on the health and nutrition of the sheep during each year of fleece growth. One serious illness can cause tender, brittle wool, a weak portion in every fiber of the whole fleece.

The beauty, luster, elasticity and strength of the wool will suffer if the sheep's diet is deficient in protein, vitamins and minerals. Mixed grain rations usually have protein content marked on the labels, and feed stores stock protein blocks, especially for use if grain is not being fed. Pasture, grain and hay provide vitamins, and hay from sunny areas is reported to have a higher vitamin content. Vitamin supplements are available, especially for older ewes.

Minerals can be supplied in a variety of ways. Trace minerals can be mixed with salt, or obtained in a trace mineral salt block.* You seldom know what mineral is actually low or missing in the soil of your locality, so some type of mineral additive is good. Lack of minerals causes specific ills, and in sheep this is directly noticeable in the wool.

WOOL GRADING: "COUNT" OR "BLOOD"

Some farm flocks specialize in wool-type sheep, others primarily in meat-type, but generally the trend is toward all-purpose breeds, fairly good on both meat and wool production.

The "blood" grading and the "count" system of wool classing does not much affect the small flock owner with no great amount of wool to sell, but you may find the explanation of interest.

There are two methods of designating the grade of wool. One of these is "spinning count" and it originally meant that one pound of fleece wool of a

---

*Salt blocks are more convenient, but are reputed to be harder on sheep's teeth than loose salt.

Clean silver-gray Romney fleece. This is the palest shade of black-sheep's wool.

"BLOOD" AND "COUNT" SYSTEMS COMPARED

"Blood" System Grades	*"Count" System*	*Example*
Fine	64s, 70s, 80s	Merino Rambouillet (some)
½ Blood	60s – 62s	Rambouillet Romeldale (some)
⅜ Blood	56s – 58s	Suffolk Southdown Corriedale
¼ Blood	48s – 50s	Oxford Dorset Romney
Low ¼ Blood	46s	Romney Oxford
Common and Braid	36s – 40s – 44s	Leicester Lincoln Cotswold

particular designation would spin that many "hanks" of wool, a hank being 560 yards. So, 70s would spin 70 hanks, 60s would spin 60 hanks. The count system usually went only as fine as 80s count, but German Saxony Merino has been known to grade 90s, where one ounce of the single fibers laid end to end, would stretch 100 miles! Count is used more in foreign countries than here, and is always expressed in even numbers.

The other way is the "blood" system of grading the fineness of wool, and it originally indicated what fraction of the blood of the sheep was from the Merino breed, which produced the finest diameter wool. This term no longer relates to Merino or part-Merino blood in the sheep, but qualifies the degrees of fiber diameter.

The accompanying chart shows the relationship of the "blood" system and the "count" system, giving examples of each. Examples are approximate, as there is variation within most breeds.

LUMPY WOOL

One disease that can destroy the fleece is *mycotic dermatitis*, a chronic skin infection which mats the wool and makes it hard to shear, and of little value. It also predisposes the sheep to fleeceworms (maggots). A well-nourished sheep with a dense fleece is not so prone to this infection.

Treatment.   Injections of penicillin-streptomycin, and a dipping or spraying with a 0.2 percent solution of zinc sulfate are useful in checking the spread of the infection.

WOOL FOLLICLE DENSITY IN VARIOUS BREEDS

Sheep Breed	Average number of follicles per square inch	Sheep Breed	Average number of follicles per square inch
Merino (fine)	36,800–56,100	Border Leicester	10,300
Merino (medium wool)	36,100–51,600	English Leicester	9,290
Merino (strong wool)	34,200–41,900	Lincoln	9,420
Polwarth	28,400–34,800	Swedish Landrace (fine)	9,350
Corriedale	14,800–19,400	Swedish Landrace (carpet)	8,260
Southdown	18,100	Cheviot	9,420
Dorset Horn	11,900	Welsh Mountain	8,900
Ryeland	10,300	Scottish Blackface	4,520
Suffolk	13,200	Wiltshire	7,350
Romney Marsh	14,200		

H. B. Carter, *Animal Breeders Abstract*, 1955.

CHAPTER EIGHTEEN

MEAT AND MUTTONBURGER

Don't hesitate to put mutton in your freezer because you fear it may be tough and inedible. The leg-of-mutton makes a wonderful "smoked ham," and most of the rest can be trimmed and boned out for use as ground meat. I have assembled here a few tested "muttonburger" recipes that give make-ahead meals, casseroles, quick-fix recipes, large recipes that make one for now and several to freeze, and some sausage recipes too. You will find that the money you can get for your culls is not nearly as much as they are worth when you put them in your freezer.

To assure more tenderness of either lamb or mutton, have your slaughterhouse use the "Tenderstretch" hanging method.

TENDERSTRETCH

Texas A & M University developed a method of carcass hanging which improves the tenderness of most of the larger and important muscles of the loin and round (most of the steaks and roasts). It is called "Tenderstretch," and consists of suspending the carcass from the aitchbone, within an hour of slaughter. The trolley hook should be sterilized before inserting in the aitchbone on the kill floor.

This method does not require any change in equipment in small slaughterhouses, and is also suitable for farm use.

It prevents the shortening of the muscle fibers as the carcass passes into rigor mortis. Before that, the muscles are soft and pliable, and if cooked rapidly are very tender. But after rigor mortis has developed, the shortened muscles become fixed and rigid. It takes an aging period of seven to fourteen days at temperature between twenty-eight and thirty-four degrees before the muscles lose their rigidity and become more pliable.

With Tenderstretch hanging right after slaughter, meat is as tender in twenty-four hours of chilling as if it were aged, and a further aging will improve the tenderness even more. Thus, with little extra effort, and no additional cost, there is a great improvement in the tenderness of many of

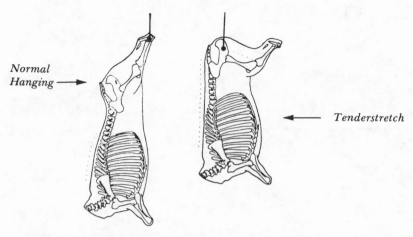

Normal Hanging ⟶

⟵ *Tenderstretch*

This shows the normal (left) and the Texas A & M "Tenderstretch" method (right) of hanging of a carcass. Points of hanging are the Achilles tendon, in left illustration, and the obturator foramen of the aitch bone, at right.

the important cuts in the animal. It does not produce the mushy over-tenderness which sometimes results with enzyme-tenderized meat.

CUTTING INSTRUCTIONS FOR MUTTON

To get the maximum use and enjoyment of a mutton carcass, give these instructions:

1. Cut off lower part of hind legs for soup bones.
2. Have both hind legs smoked, for leg-of-mutton "hams."
3. Package riblets (spare ribs) and breast in two-pound packages, to pressure cook and separate meat, for curry recipes. These parts are hard to bone out.
4. Have tenderloin removed and made into boneless cutlets.
5. Have the rest boned out, fat trimmed off, and ground. Double wrap in one-pound packages and try my muttonburger recipes in this chapter. Do not be surprised when the ground mutton seems a lot juicier than other ground meat. Older animals' tissue can bind large amounts of water.

CUTTING PLAN FOR LAMB

This slightly different plan of cutting a lamb is simple, and not far from the standard cutting procedure. It was the suggestion of Mrs. Nancy Wetherbee, and was shown in *Shepherd* magazine in 1968.

The rib on the chops is cut very short. This leaves no uneaten waste,

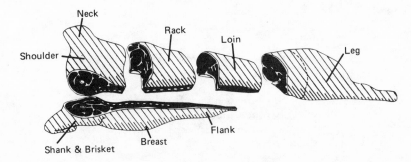

Neck

Rack

Loin

Leg

Shoulder

Flank

Breast

Shank & Brisket

There are several correct ways to break a lamb carcass, and no one method can be considered best. However, for many purposes, the method shown is ideal. (Lamb Cutting Manual, published by American Lamb Council and National Livestock and Meat Board.)

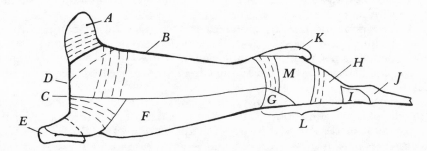

Here's a detailed method of cutting a carcass, suggesting uses for the individual cuts. (A) Neck: slice for braising or bone for stew or ground lamb. (B) Loin, rack, shoulder: cut for roasts or cut between ribs for chops. (C) The lateral cutting of the carcass is at a high level, leaving less rib ends on the chops and more in the spare ribs. (D) Lift shoulder blade for roast; bone if desired. (E) Braise trotters or cut for soup. (F) Breast: cut between ribs for braising. (G) Grind or cut for stew. (H) Center-cut steaks. (I) Braise trotters. (J) Discard or use for soup. (K) Discard or use for soup. (L) Leg. (M) Optional removal of sirloin steaks from leg roast. (Shepherd magazine)

makes a much nicer chop and longer spare ribs. They can be braised, or pressure-cooked and all the lean meat separated from the bone and fat, for a delicious curry. (See recipe in this chapter.)

With this type of cutting, the larger the lamb the better, if young and not overly fat. If the leg is too large, remove a few more chops and steaks, either from the center of the leg or from the loin end.

RECIPES

SMOKED LEG OF MUTTON "HAM"

Glaze: ½ cup brown sugar, firmly packed
1 teaspoon mustard
1 cup orange or pineapple juice
Cloves

Method: Soak mutton in cold water for one hour. Dry with paper towels and wrap securely in a large piece of aluminum foil. Seal edges well and place in a baking dish. Bake at 350° for 30 minutes to the pound.

Mix together brown sugar, mustard and fruit juice. Place the precooked leg in a baking dish. Score outer covering with knife and pour juice mixture over it. Stud with cloves. Bake the leg for an additional 40 minutes, basting often with pan juices. Serve hot or cold.

Note: Simmering may be preferred for the first stage. Soak mutton as above. Plunge into large pan with warm water. Bring to a boil and simmer for 30 minutes per pound, or until tender. Allow leg to cool in the liquid. Drain and refrigerate covered (do not freeze) until needed, then bake with glaze.

In our experience, the yearling or two-year-old is very tender and tasty. But the old, old ewe is tasty and not very tender. So, cut the real old ham into several pieces that will fit into your pressure cooker. Do the one-hour soak, then pressure cook for 15 minutes at 15 pounds of pressure. I then bake the "ham." It will not need a long baking and will be both tasty and tender. Grind up leftovers for ham hash. Cook split peas with the bone.

Mutton ham recipe is from the Australian Meat Board, as printed in *Shepherd* magazine in 1973.

MUTTON PATTIES FOR THE FREEZER

5 pounds lean ground mutton
5 tablespoons water
4 tablespoons lamb seasoning salt
1 teaspoon coarse ground pepper
1 teaspoon crushed mint leaves
1 teaspoon fenugreek (optional)
1 beaten egg
1 tablespoon wheat germ

Mix all together. Press into patties with hamburger press (or divide in about ¼ pound balls and press flat). Package for freezer in desired number of servings to the package, separated by a double thickness of wax paper or foil. Double wrap in plastic or freezer paper.

To serve, thaw for 45 minutes in the package and then separate, or separate and thaw for 25 minutes.

Broil, pan fry, or bake in gravy.

Seasoning Salt for Lamb and Mutton

> 1 cup fine plain popcorn salt (dissolves instantly,
>   doesn't fall off the meat)
> 1 teaspoon black pepper
> 1 teaspoon paprika
> ½ teaspoon ground ginger
> ½ teaspoon dry mustard
> ½ teaspoon poultry seasoning
> ⅓ teaspoon cayenne pepper
> 3 teaspoons garlic salt (or 2 teaspoons garlic powder)

Combine all ingredients. Mix well and pack in shaker jars.

This recipe originated at Purdue University's Home Economics Department, and was reprinted in *Shepherd* magazine.

BREAKFAST SAUSAGE

> 1 pound lean ground lamb or mutton
> ⅛ teaspoon coarse ground pepper
> ½ teaspoon salt (or more)
> ¼ teaspoon powdered marjoram
> ¼ teaspoon powdered thyme
> ¼ teaspoon powdered sage (or more)
> ¼ teaspoon savory

Mix all ingredients together thoroughly. Cover bowl, place in refrigerator overnight. To use, shape into patties about ½ inch thick. Cook over moderate heat in heavy skillet until brown. Turn. Brown other side, lower heat to cook through. Serves 5–6.

If you like your sausage a little more moist, you can add about 2 tablespoons water and cover the skillet, when you lower the heat to cook. For a larger quantity to freeze in rolls for slicing later, add a little ice water and mix in with sausage, so it doesn't crumble when you defrost and cut it into slices. We like a little more salt than in this recipe, and a lot more sage.

This recipe was in *Shepherd* magazine, April, 1972.

LAMB SCOTCH ROAST AND CHOPS

The lamb Scotch roast and chops are both excellent ways to use the lamb breast section and lean trimmings all at one time. Square off the breast by removing the brisket point and flank section. Run knife along the top of the rib bones and lift the top cover. Leave the top cover attached at the brisket side. Stuff the breast section with lean ground lamb, and either tie or net the Scotch roast.

For Scotch chops, let the cut firm up in the cooler, then cut in between each rib. (National Livestock and Meat Board)

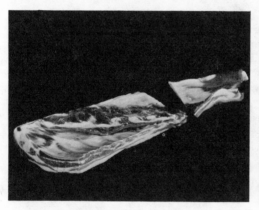

1. *Square up the lamb breast section.*

2. *Lift the top cover.*

3. *Stuff with lean ground lamb.*

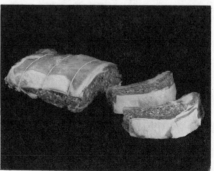

4. *Economy Scotch roast and chops.*

HASTY HASH

1 pound ground lamb or mutton
1 tablespoon oil
1 small onion, chopped
½ teaspoon salt
⅛ teaspoon garlic powder
½ teaspoon freshly ground pepper
4 tablespoons soy sauce
2 cups raw potato, shredded
   (can be defrosted frozen hashbrowns)

Sauté meat with oil until pink color leaves, add onions, sauté until onions are transparent. Separate meat with a fork, as it cooks. Stir in salt, garlic powder, pepper, and soy sauce. Mix. Layer potatoes on top of meat, cover pan, and cook on medium-low heat for 20 minutes, stirring gently from time to time. Uncover and turn heat up a little. Stir and cook until potatoes are beginning to get brown. Good with catsup.
   Serves 4.

QUICHE PIE

1 pound ground lamb or mutton
1 tablespoon dehydrated chopped onion
4 eggs, slightly beaten
1½ cups grated cheddar cheese
½ cup milk
1 teaspoon salt
¼ teaspoon pepper
2 tablespoons fresh parsley, minced
1 unbaked 8" or 9" pastry shell
Paprika

Cook ground meat until lightly browned, adding chopped onion after meat loses its red color. Separate meat as it cooks. Drain off fat. Combine eggs, 1 cup of the cheese, milk, salt, and pepper, add meat and onion mixture, and stir together. Turn into pastry shell. Top with the rest of the cheese, sprinkle with the parsley, and dust lightly with paprika. Bake 35–40 minutes at 420°, or until firm in the center. If it starts getting too brown, cover with loose tent of foil.
   Serves 6.

Recipe from *Shepherd* magazine, January, 1972.

EASY MEAT PIE

> 1 pound lean ground lamb or mutton
> 1 large onion, chopped
> 2 beef bouillon cubes
> 2 tablespoons boiling water
> ¼ teaspoon garlic powder
> 1 can condensed cream of potato soup (10¾ ounce)
> Pastry for top and bottom crust

Sauté ground meat and onion slowly in skillet, just until meat barely loses its red color, separating with a fork. Melt bouillon cubes in boiling water, add to meat, and mix in garlic powder and undiluted soup. Salt, to taste. Pour into pastry lined 9 inch pie tin, cover with top crust. Crimp edges to seal, and slit the top crust in several places. Bake 15 minutes at 425°, reduce heat to 350° and bake 45–55 minutes longer. If top is getting too brown, cover loosely with aluminum foil.

Serves 6.

GARDEN MEAT LOAF SQUARES

> 2 tablespoons oil
> ⅔ cup chopped onions
> 1 cup fresh string beans, cut small, or drained canned beans
> ½ cup green pepper, chopped
> 1 cup celery, chopped
>
> 2 pounds ground lamb or mutton
> 1 cup bread crumbs
> 2 teaspoons salt
> ½ teaspoon fresh ground pepper
> 1 tablespoon Worcestershire sauce
> 1 teaspoon soy sauce
> 1 egg, beaten lightly
> ⅔ cup tomato juice
> Garnish, catsup or chili sauce

Sauté onion, beans, pepper, and celery in oil until tender. Mix meat, seasonings, egg, and tomato juice. Mix in the vegetables. Press mixture into a 9 x 13 x 2 inch pan. Bake ½ hour at 350°. Spread top with thin layer of catsup or chili sauce, bake 5 minutes more. Cut into squares to serve.

Serves 8. (Or serve 4, and reheat the rest for another meal. Good reheated.)

LAMB ROLLETTES WITH CURRY SAUCE

1 pound ground lean lamb or mutton
⅓ cup finely shredded carrot
½ cup fine bread crumbs
¼ cup milk
1 tablespoon minced onion
¾ teaspoon salt
¼ teaspoon ground allspice
⅛ teaspoon coarsely ground black pepper
4 slices bacon, slightly cooked, drained
   Curry sauce

Lightly mix lamb, carrots, bread crumbs, milk, onion, salt, and pepper. Form into 4 small loaves. Wrap a strip of bacon around each loaf, tucking ends under loaf. Place in shallow baking dish. Bake at 350° for 25–30 minutes, or until slightly browned. Serve hot with curry sauce.

Curry Sauce: Melt 2 tablespoons butter in saucepan. Stir in 2 tablespoons flour, ½ to 1 teaspoon curry powder and ½ teaspoon salt. Gradually stir in 1⅓ cups milk and cook, stirring, over medium heat until thickened and smooth. Add ½ cup cooked green peas. Heat through, stirring constantly, serve as sauce over lamb loaves. Serves 4.

This recipe originated with the American Lamb Council, was reprinted in *Shepherd* magazine.

ANNA'S CASSEROLE

1¼ pounds lean ground mutton
1 package (10 ounce) frozen peas, defrosted
2 cups thinly diagonal sliced celery
1 can (10¾ ounce) cream of chicken or cream of mushroom soup
1 teaspoon lamb seasoning salt, divided
½ teaspoon freshly ground pepper
1 onion, chopped, or 2 tablespoons dried onion flakes
1 small package (⅞ ounce) crushed barbeque potato chips
   Paprika

Sauté ground mutton and half of the onion in skillet until lightly browned, seasoning with half of the lamb salt, breaking apart with a fork as it cooks. Drain off fat. Spoon meat into medium size loaf pan. Scatter defrosted peas over the meat. Layer the celery on top of the peas. Mix pepper and the rest of the onion and seasoning salt with the can of soup, spread on top. Put crushed potato chips on it, and sprinkle well with paprika. Bake about ½ hour at 375°.

Serves 4.

MUTTONBURGER STROGANOFF

1 pound ground lamb or mutton
2 tablespoons dried minced onion
1 tablespoon butter or margarine
½ pound fresh mushrooms, sliced
1 can (10½ ounce) cream of chicken soup
  Salt to taste
  Pinch of nutmeg
1 cup sour cream

  Egg noodles
  Freshly ground pepper

Brown the ground meat in margarine until red color disappears, add minced onion and sliced mushrooms, sauté until tender. Add chicken soup, salt and nutmeg. Heat to simmering.

Cook noodles until tender, and drain well. Add sour cream and noodles to meat mixture, stir gently. Reheat very hot, but do not boil. Serve on heated platter, topped with fresh ground pepper.

Serves 4.

TORTILLA PIE

1½ pounds ground lamb or mutton
½ cup chopped onion
1 medium green pepper, chopped
½ teaspoon salt, or to taste
1 teaspoon coarse ground pepper
1 can (15 ounces) tomato sauce
1 cup shredded medium cheddar cheese
1½ cups crushed corn chips
6 thin slices cheddar cheese

Sauté meat, onion and green pepper until meat has lost its red color, separating with a fork as it cooks. Pour off fat. Add tomato sauce, pepper and salt to taste. In shallow casserole, put ⅓ of the corn chip crumbs, then half of the meat mixture and half of the shredded cheese. Repeat layers, then top with rest of the corn chip crumbs. Bake 30 minutes at 350°, then top with the sliced cheese and bake about 5 minutes more. Good served with lettuce salad.

Serves 4–6.

TOP-OF-THE-STOVE MEAT LOAF

$1\frac{1}{2}$ pounds lean ground lamb or mutton
$1\frac{1}{4}$ cups quick cooking oats
   1 cup ground and drained (reserve juice) green tomatoes
   1 teaspoon salt
$\frac{1}{4}$ teaspoon garlic powder
$\frac{1}{4}$ teaspoon freshly ground pepper
   1 small onion, chopped fine
   1 tablespoon chopped parsley
   1 egg, slightly beaten
   1 cup green tomato juice, plus enough tomato juice
     to make the cup
   2 tablespoons vegetable oil (less if you use teflon or
     T-Fal skillet)
   Sautéed onion rings

Combine all ingredients except oil and onion rings. Shape into a large rounded patty, slightly smaller than large skillet. Pour oil into hot skillet, place meat patty in hot oil, and cover tightly. Cook over medium heat for 15 minutes. Loosen with spatula and turn patty over, browned side up. Cut into 6 wedges, separating them slightly. Cover and cook 15 minutes, or until done. Serve topped with sautéed onion rings.
   Serves 6.

Recipe from *The Green Tomato Cookbook* by Paula Simmons, published by Pacific Search.

MEAT BALLS IN SPAGHETTI SAUCE

   1 pound ground lamb or mutton
   1 cup grated cucumber (or zucchini)
$\frac{1}{2}$ cup dry bread crumbs
   1 egg, beaten
   1 teaspoon salt
   Dash cayenne pepper
$\frac{1}{2}$ teaspoon Tabasco sauce
   1 tablespoon dried onion flakes
   1 can ($10\frac{1}{4}$ ounce) spaghetti sauce (meatless)

Combine all ingredients except spaghetti sauce. Form into small balls, cook over low heat until browned on all sides. Add spaghetti sauce. Cook over low heat about 20 minutes, serve over spaghetti, with Parmesan cheese.
   Serves 4.

Recipe from *Shepherd* magazine, June, 1966.

MOUSSAKA CASSEROLE

1 pound ground lamb
2 chopped onions
4 chopped tomatoes (or one 14½-ounce
    can tomatoes, drained)
¼ cup red wine
⅛ teaspoon freshly ground pepper
⅛ teaspoon cinnamon
3 tablespoons chopped parsley
  Pinch thyme
  Salt, to taste
1 medium eggplant, peeled and cut
    into ½-inch thick slices
2 medium zucchini, cut in ½-inch thick slices
  Vegetable oil
1 cup cottage cheese, drained (or dry curd)
1½ cups white sauce, medium thick
½ cup soft bread crumbs
¾ cup grated Parmesan cheese

Heat large skillet; sprinkle with salt; add meat and onions; stir-fry until
meat starts to brown. Add tomatoes, wine, pepper, cinnamon, parsley, and
thyme. Simmer until thick and liquid has cooked out, stirring occasionally. Add salt to taste. Sauté eggplant and zucchini slices in oil until almost
tender. Add cottage cheese to white sauce. In buttered 9 x 13 inch pan or
large casserole, layer the eggplant on the bottom. Combine bread crumbs
and Parmesan; sprinkle half of it on the eggplant; top with half meat mixture. On top of this, layer zucchini slices; sprinkle with rest of bread crumbs
and Parmesan. Add rest of meat mixture. Top with white sauce. Bake at
350° for one hour or until bubbly and browned. Let casserole stand for 15
minutes or more before serving. Serves 6.

From *The Zucchini Cookbook* by Paula Simmons, published by Pacific Search.

CHAPTER NINETEEN

PROFIT FROM SHEEP

The key to profit is to make good use of all your potential sources of income, connected with your sheep. This requires good planning and good management, so that all your ewes will lamb, with a high percentage of twins, and low percentage of lamb deaths. Cull out your poor producers, and replace them by keeping the best of your early-born, fast-growing twin ewe lambs.

Having heavy fleeces on your ewes and your ram will give added pounds of wool to sell. Doing your own shearing keeps expenses down. Keep your wool in good condition, and market it at the best price per pound.

Control parasites, so you are feeding sheep, not worms, and so the sheep are in good condition, which makes them more resistant to disease.

Publicize whatever superiority your breed has, to make money by selling breeding stock.

For both profit and pleasure, make use of all the by-products that you can.

SOURCES OF INCOME

1. Locker lambs
2. Mutton
3. Breeding stock
4. Pelts
5. Shearing
6. Manure for gardens
7. Soap
8. Candles
9. Building of sheep "furniture"
10. Waste wool for insulation
11. Wool to spinners
12. Black sheep
13. Handspun yarn
14. Incentive payments

LOCKER LAMBS

If you are raising only a few sheep, you will find that the meat lambs can be sold at a much better price by dealing directly with the customer. At a better price than you would get from a "lamb pool," you can still be giving them a better buy than they would get at the meat market.

195

Most states have some restrictions on slaughter-and-sell practices, some designed to deter rustling, others to enforce sanitation. In our state, if a lamb is taken to the slaughterhouse for inspection and butchering, and cut and wrapped, each package of frozen meat must be stamped "not for sale." However, this restriction on private slaughterhouses need not stop you from legally selling locker lambs.

Sell to your customer in advance, deliver the lamb to the slaughterhouse and give them your customer's name. They will notify you, or him, of the cutting weight as you direct them. Collect the price per pound on that weight from your customer, who then picks up his meat from the slaughterhouse, all cut and wrapped and frozen, and pays them for the cut-and-wrap charges. Normally, the seller of the lamb pays for the slaughter charge, which is a nominal flat fee per lamb.

Taking orders in advance is always a good idea, not just to stay legal while selling locker meat, but so you have your whole crop sold, and can deliver it about the time the summer pasture starts to dry up. Fast growth of your lambs will assure that they are ready for marketing by then.

Fast growth is also associated with tenderness, so if a lamb takes longer getting to locker size, it may not be quite as tender as if it had a rapid growth rate.

Young lamb is naturally expected to be tender but several factors can, one at a time or combined, work against this tenderness:

1. Stress imposed on animals prior to slaughter, such as rough handling when catching and loading.
2. Failure to age the carcass long enough in the chill room. One week is not too long, when hung by the standard method.*
3. Slow growth rate. This is a good reason to grain your lambs in a creep feeder.
4. Drying out in slow freezing. Most cut-and-wrap facilities will do the freezing, and faster than it could be done in your home freezer.
5. Length of time in freezer storage. One year should be the maximum storage.

Organic lambs. With the current trend to health consciousness, there is a very special market for organically-grown locker lamb, which can only be met by the small grower. Big feeders and producers have more of a disease problem than is ordinarily present on a small farm, so have to use vaccinations and injections and medicated food as a preventive measure, even when disease is not present.

Very strict organic raising is difficult. You must provide unmedicated feeds, which means whole and crushed grains, not pelleted feeds that contain additives. If you buy your grain direct from its grower, you will know

---

*For added tenderness by a new hanging procedure, see "Tenderstretch" described in Chapter 18 on Muttonburger.

whether it has been sprayed or treated in any way. In this matter, a small grain producer is less apt to use a lot of chemicals, mainly because their use is only profitable on a large scale.

Extremely good sanitation practices can make medications less necessary. You can avoid chemical dipping for ticks if you once get rid of ticks completely, and do not allow them to be brought in by a strange sheep.* We loan out our rams for breeding services, and have to de-tick them when they come home, or have ticks spread to all the ewes.

You should worm your ewes regularly, and then keep the lambs in clean surroundings with creep feed and "advance creep" pasture feeding, where they get into each fresh pasture ahead of the ewes. When this is done, there is a very good chance that you will be successful in preventing a worm build-up in the lambs that would necessitate worming them before locker age.

In general, lamb consumers are among the higher income groups of the population, so this should determine where to advertise, if advertising is needed to sell your locker lambs.

Lambs for Easter. Creep feed your lambs and try to have some of them ready for sale by Easter. The eating of lamb is part of the religious festivities in the Greek Orthodox tradition, among others. If you have lambs born early (first half of January or before) and do not have them promised, you might tell the nearest Greek Orthodox church of their availability, or advertise if there is a Greek newspaper in your area. The size preferred in the Northeast is about 35 to 40 pounds liveweight; in the West the ideal size is a little larger. Lambs sold at that size are called "milk-fed." The term "hot-house lamb" is sometimes applied to the early January lambs that are sold at Easter, and sometimes to the fall lambs, born out-of-season and raised mostly indoors, for early spring sale.

Walk-in refrigerator. Utah State College engineers, Logan, Utah, have plans for a large walk-in refrigerator that you can build. This would make on-the-farm processing a practical alternative to the slaughterhouse, and could also be used for vegetables from a home garden. The four-page blueprints cost about $4, and are a complete construction guide.

MUTTON

Selling an aging ewe or an extra ram is not as easy as selling a lamb, which is expected to be more delicate and tender, for mutton has a rather bad image in this country. Many people (even people who have never tasted it)

---

*Although not officially "approved" for sheep, rotenone is the one organic dip that is the choice of organic growers when they find dipping necessary. See Chapter 14 on External Parasites.

say they don't like it, and expect it to be tough and strong in taste. Preju-
dices are hard to overcome, so consider saving the mutton for your own
locker. You will be pleasantly surprised to find that there are a lot of uses
for mutton so that you will be able to utilize whatever "culls" you have, and
enjoy doing it.

Keep in mind the known digestibility of mutton, which makes it a good
meat for people who have various digestive difficulties.

A grain-fed farm ewe or ram makes good eating. For the person who
thinks he doesn't like it, or expects it to be tough, see the previous chapter
of recipes.

PUREBRED OR SPECIALTY BREEDING STOCK

In raising purebred and/or registered sheep on a small scale, some profits
should be made in the sale of breeding stock. If they are not, you have lost
some of the advantage of paying for purebred and registered sheep, for the
actual receipts for sale of wool and meat would be very little different with
less expensive breeding stock. To raise registered sheep successfully involves
a tremendous amount of record-keeping and either experience with sheep
or good planning in order to improve the flock or just to keep it from
deteriorating. Beginners are often advised to start with less expensive sheep,
so there is less money involved in losing a sheep due to inexperience.

If you are buying purebreds and plan to sell them, try to select a breed
that would appeal to a market that you are familiar with, if possible, as
well as one suitable for your area. Some unusual breeds are in demand for
non-commercial raising, with good sales of breeding stock. Some breeds
thrive at high altitudes, some do well in heat and some prefer cooler
climates, some graze well on rolling hills, and some are more at home on
flat meadows.

For instance, if we were going to raise purebred registered stock, we
would probably choose Romneys. They have a good fleece for spinning,
and our wool customers would be spinners. They are a good meat animal
for our locker lamb sales. Most of all, they are especially well adapted for
raising in our very wet climate.

In actual practice, we are not raising pure Romneys, for we have a more
specialized field. We are specifically raising "black sheep," and we use their
wool in our own handspinning business, although we could easily sell all of
our fleeces if we did not want them for our own use. The breeding stock is
valuable because of the handcraft demand for dark wools. Our sheep are
mostly crosses of Romney, Lincoln and Columbia. Some are mostly Lincoln,
and most are a part-Lincoln cross, producing fleeces of various types, all
suitable for our own yarn spinning. While this wet climate is not ideal for
Lincolns, we need their wool for some special uses, and like the part-
Lincoln lambswool.

If raising black sheep appeals to you, see the section on them, later in
this chapter.

PELTS

The pelts of meat lambs can be another source of income. If you are a spinner, you may prefer to shear the lamb just before it goes to the locker, or if you want both the fleece and the pelt, you can shear it and then wait about six weeks before it is slaughtered. This way you will have enough wool on the skin to use as a "shearling," for slippers or jacket lining.

For tanning, skins should not be damaged by ticks, which is another reason to keep your sheep free of them. The dark lumps caused by tick bites are called "cockle" in pelts or leather. If you use tanned shearling pelts to make jackets with the wool inside, the outer surface can be sanded to produce a beautiful suede finish. However, cockle defects would seriously impair the softness and appearance of the sueded leather.

Shearing nicks will also show up in the pelt. Skinning should be done carefully, to avoid cuts into the hide.

Most slaughterhouses realize a small income from the sale of pelts they get from the animals being processed, but you can ask for the pelt back. Pick it up the day after it is skinned.

Pelts should be liberally salted on the flesh side, as soon as possible after the body heat has left them, to avoid spoilage. If you are going to start home tanning immediately, spoilage is no problem. If you are sending them away for tanning or will not be tanning until some weeks or months later, the salting will preserve them until they are tanned.

Shop around for custom-tanning prices, for often you can get a better price when sending them away, even with paying postage both ways. See Sources chapter for mail-order tanning.

You can get catalog supplies for tanning, and also kits that supply everything that is needed. To decide about the price of home tanning versus tannery tanning, estimate the cost of your materials, and also the value of your time if you have little to spare. Weigh this against the cost of postage and tanning.

When trying any tanning process for the first time, be cautious and do only one pelt. When you have done it once, you may see ways to do a better job than you did the first time, or may prefer to try one of the other processes, to see if it is easier and more satisfactory.

Once you perfect your system of tanning and have done it a few times, you should find a ready market in local craft shops or even decorator shops. To get a better price you can sell direct to your customers.

Three ways of tanning, in pamphlet. *Shepherd* magazine has written articles on three ways of tanning sheep pelts, including the USDA's glutaraldehyde tanning process, well illustrated. This fairly new process produces machine washable pelts that are becoming very popular in hospitals and rest homes as bedpads for invalids. The pelts distribute pressure evenly, dissipate moisture, do not wrinkle or chafe, and prevent ulcers and bedsores.

While the tanning chemical is dangerous to handle, the results can be worth the trouble. Washable pelts are also good for car seat covers, rugs, and sewn into jackets. Instructions for this, for the urea-formaldehyde tanning and the salt-alum tanning, are all in reprint form, listed in the Appendix.

In addition to these three ways of home tanning, there is also the salt-acid process, which I will describe. As with most home tanning processes (except glutaraldehyde), the resulting pelt is not washable, but it can be cleaned with gasoline-and-sawdust. Also, the acid must be handled cautiously, and neutralized carefully so that it does not remain on the skin and damage it.

Fleshing out the pelt. First, scrape the flesh side with a heavy knife to remove all meat, tissue and grease. Do not injure the true skin, or expose the hair roots. Scrape off all tough membranes and inner muscular fleshy coat.

Salting the pelt. If you are not going to tan the skin the day it comes off the sheep, you should salt it heavily to preserve it for later tanning. As soon as the animal heat has left the pelt, rub common pickling salt into the flesh side. A lamb pelt will take about three pounds. Do a thorough job, giving attention to salting the edges well. Spread the pelt out to dry, flesh side up.

Preparing *salted* pelt for tanning. Later, when you want to prepare this pelt for tanning, soak it overnight in a large tub of cold water containing one cup of laundry detergent and one cup of pine-oil type disinfectant. In the morning, remove this water by spinning the pelt in the "spin" cycle of your washer. Then wash the pelt in the washing machine, short cycle with cool water and laundry detergent. Rinse. Spin the rinse water out in the "spin" cycle, then proceed with tanning.

Preparing a *fresh* pelt for tanning. First, flesh out the skin as directed above. Then sprinkle the skin side of the pelt with strong detergent, and brush it with a stiff brush.

Wash the pelt in the washing machine, short cycle with cool or lukewarm water and detergent. Rinse. Spin out the water, using the "spin" cycle of the washing machine, and proceed with your choice of tanning processes. All the fat, blood and dirt should be removed from the pelt by now.

Salt-acid tanning process. For the salt-acid tanning solution, use a plastic drum or plastic garbage can. Metal containers must not be used. For best results, solution must remain at about room temperature — between 65 and 75 degrees.

Solution: For each one gallon of clear 70° water, use one pound pickling and canning salt and *one* of the following acids: 1 oz. concentrated sulphuric acid, or 4 oz. new Battery Fluid (acid) or ½ cup sodium bisulfate, dry crystals or 2 oz. oxalic acid crystals.

Use your choice of only *one* of the above acids, with the water and salt, for tanning. A choice of acids is given so that you can use the one most easily obtained in your area. Whichever acid you use, measure it out carefully, and store the acid in a safe place. If you are measuring liquid acid, use a glass or plastic cup, not metal. Add it *slowly* to the water, letting the acid enter at the edge of the water. Rinse the measuring cup in the solution, and stir the mixture with a wooden paddle.

Immerse the pelt in the tanning solution, push it down with the wooden paddle, and stir slowly. Leave the pelt in the solution for five days (or more, up to two weeks if the solution does not get over 75 degrees). Keep the pelt submerged, and stir it gently from time to time.

Neutralize the tanning solution. Remove the pelt and spin out the tanning solution in the "spin" cycle of your washer. Rinse the pelt in clear water twice, then spin out the rinse water. Immerse the pelt in a solution of water and borax, using one ounce borax to each gallon of water. Work the pelt for about an hour in this, then rinse out in clear water. "Spin" out the rinse water. This step is necessary to neutralize the acid solution, so that it does not remain on the skin and damage it.

Tack the pelt out flat, flesh side up. Apply a thin coat of neatsfoot oil to that side. While the oil is soaking in, taking from eight to ten hours in a warm room, you can dry out the wool side, using a fan or a hair dryer. Then apply a thin coat of tanning oil or leather dressing on the flesh side.

Drying and softening the pelt. When the tanning oil has soaked in, allow the pelt to dry until it starts showing light colored places. Remove it from the frame, and start the softening process. Stretch the skin in all directions, and, flesh side down, work it over the board, to soften the skin as it finishes drying.

You can sandpaper the flesh side when dry, to make it smooth. Comb out the wool with the coarse teeth of a metal dog comb, and finish with finer teeth. If the wool seems too fuzzy and dried out, you can rub a hairdressing (such as VO5) on your hands, and rub them lightly through the wool, then brush it gently. Repeat if necessary.

LOCAL SHEARING

If you learn to shear your own sheep, you will not only save the price of shearing, but have a skill that can be used to bring in a part-time income. In many areas, shearers are scarce and sheep raisers have to wait until the heat of the summer before they can hire one. If they happen to have only four or six sheep, most commercial shearers don't want to spend the time to travel some distance for the small fee that could be charged. Another reason a professional shearer would not want to do a small number of sheep is that facilities are seldom ideal. Often there is no good method of catching the sheep, and there is no electricity for his shearing equipment.

When you shear with hand shears, which are so convenient for a small number of sheep, you need not worry about electricity. And if you are shearing in your own vicinity you will not travel such a distance, and can have an arrangement made ahead that the owner will have the sheep penned at the time you arrive. When there are only a few sheep, you can often trade shearing services for the actual wool that you get from their sheep. They would hardly have enough wool to sell at a good price if they were to pay you and keep their wool. If they want to keep their wool for their own use in spinning, they will be willing to pay you a fair price for shearing, and expect you to shear the fleece carefully.

When you shear "for the wool," they will want you to trim the hooves and worm the sheep as part of your service. The wool that you get could be sold along with yours to provide income, or if you are a spinner, this would be more wool for your spinning projects.

MANURE

Another potential income is from the sheep manure, either selling it or using it in your own garden. It not only stimulates the crop growth, but adds valuable humus to the soil, which is not true of chemical fertilizers. You don't have to be modest about proclaiming its superiority over that of other animals, as the accompanying USDA chart will show.

Pounds per ton of:	Nitrogen	Phosphorus	Potash
Sheep manure	20	9	17
Horse manure	11	6	13
Cow manure	9	6	8

As sheep make use of ingested sulfur compounds to produce wool, their manure does not have the unpleasant smelling sulfides found in cow manure. It is also in separate pellets, or in pellets that hold together in a clump, is less messy in the garden, and does not even need aging. If you gather it for your own garden, take it first from paths and places where it will not help to fertilize the pasture. Since it contains many of the valuable elements taken from the soil by the plants eaten by the sheep, it is convenient that they spread a lot of it on the pasture. Its pelleted form causes it to fall in the grass instead of lying on top of it where it might smother the vegetation.

For use in your own garden, clean out the barn twice a year, in spring and fall. The wasted hay and bedding left on the barn floor will absorb much of the manure, containing valuable nutrients. Being inside they are undamaged by rain and sunshine, just waiting to be reclaimed. We spread a thick barn-cleaning mulch on a portion of our garden and don't even dig it in, just set out our tomato, zucchini and cabbage plants in holes in the mulch, where they grow without weeding.

HOW MUCH VACCINE?

				½cc	Tommy + Penicillin	DIED 2-27		
BO-SE ? NEWBORN								
EWE								
TETENAS ? NEWBORN								
EWE								
OTHER SHOTS ? NEWBORN								
(PENICELLIN) EWE								
FEEDING ?								
DOCKING ?								
WEANING ?								
SIGNS OF LAMBING :								
OFF FEED A COUPLE DAYS								
SLOWER - QUIETER								
HOT PINK UDDER								
LABOR :								
½ - 2 HOURS NORMAL								
WATER SACK THEN								
2 FEET TOES DOWN HEAD AT KNEES								
CLEAN NOSE + HELP DRY								
LAMB TO PREVENT PNEUMONIA								

½-1 t WHISKY TO 1-2 oz MILK

						LAMB GOT
	2/22	TINKERBELL HAD LAMB BY HERSELF				3 FEEDING
1	2/23	(T) REFUSES TO LET LAMB NURSE				2 FEEDING
2	2/24	LAMB EATS BETTER BUT NOT GOOD ENOUGH			MUST	
		KEEP WARM + FEED 2-3 HOURS				
3	2/25	8 OZ				
	2/26	12 + OZ				
	2/27	TOMMY DIED				

SUGGESTED RATE

ACTUAL RATE

OUNCES

30
25
20
15
10
5
0

1 2 3 4 5 6 7 8 9 10 11

HOME-MADE SOAP

Home-made soap is one of the "good things" of life—and you can make a lot of it with the fat from lamb or mutton, when it is trimmed for locker packaging. Have the slaughterhouse save all the fat trimmings. Some places will grind them for you, which makes the rendering easier.

Render the tallow. Cut up chunks of lamb or mutton fat (tallow), put it in a large kettle and cook it slowly over low heat. It will take several hours for a large batch, but don't rush or you will risk burning it. When the tallow is all pretty well melted down, strain it through a cloth.

Purify the tallow. Boil the fat that you rendered, with about twice its volume of water. Strain it and set it aside to cool. The clean fat will rise into a solid block. When it has cooled and hardened, remove from the water, turn upside down, cut in wedges, and scrape off the residue of impurities from the bottom. This purified tallow will keep for several weeks in the refrigerator.

SOPHIA BLOCK'S LAMB TALLOW
SOAP RECIPE

Measure six pounds of clean purified tallow, which is about 6¾ pints of liquid tallow. Heat it slowly in a large enamel pan to between 100° and 110°.

Put 2½ pints (5 cups) of water in a smaller enamel pan. Put the pan on a protected surface. Stand back and slowly pour in one newly opened can of lye (this must be lye, not Drano). Turn your face away so you do not breathe its caustic fumes. The lye will heat up the water. Allow it to cool to 98°–100°. Use a candy thermometer, suspended from the side of the pan, not touching the bottom of the pan. When the lye is at the proper temperature, pour it into a half gallon (magnum) liquor bottle, using an agate funnel. Now put the opening of this bottle on the rim of the pot of tallow, and pour the lye mixture very slowly in a thin stream, while stirring the fat and lye together slowly and gently. It is easier if you have a helper to pour in the lye. The tallow should be at the right temperature (100°–110°) and the lye poured into it in a very thin stream. Stirring must be done slowly and very gently and steadily. If the lye is poured in too fast, or the stirring is not slow and gentle, the soap will separate or curdle and you will ruin the whole batch. Stir slowly for twenty minutes, and then pour it into prepared containers.

Soap containers. Agate photo-development pans are ideal for soap. Or use wooden boxes lined with brown paper or with clean cotton cloth, wet down with water and wrung out. Have the paper or cloth folded out over the outside edge, to make the soap easy to remove when you are ready. You can use cardboard boxes, lined with plastic wrap, which is turned back over the outside edges and stapled to hold it in place while you are pouring the soap.

Pour the soap into these prepared containers, then cover the soap with a board or heavy cardboard, and then with a blanket. This keeps it from cooling too fast. Allow it to cool and harden for a day or two in a warm place away from drafts. The soap will begin to lose its sheen as it hardens. After two or three days and before it gets too hard, you can remove it from the boxes. Cut it into separate bars to age for several weeks, or months, before use. It can be cut neatly with a fine taut wire wrapped around it and pulled tight. Age these bars unwrapped, with air circulating around them, for several weeks. Look for any liquid-appearing substance on or in the soap. That would be free lye, and you should discard the soap or re-process it.

Soap variations. Mutton tallow soap is often called saddle soap because it cleans and preserves leather so well. It can be used equally well as a bath, laundry or dishwashing soap, but by a few variations you can make it even more suited to different uses.

Perfumed soap: Add oil of lemon, oil of lavender, or other oil-perfumes (not any containing alcohol) or boil up leaves of rose geranium and use this "tea" as part of the cold water used with the lye. Reserve part of the lye-dissolving quantity of water, boil up the perfuming leaves in it, and add it to the dissolved lye when it has cooled a little. Since soap will absorb odors, it can be perfumed easily after it is in bars and aged, by wrapping it in tissue that has been wet with perfume and dried out.

Green soap can be made with "vegetable" coloring obtained by pounding out a few drops of juice from beet tops, or use the vegetable coloring sold for baking.

Mint soap: Use one cup less water to dissolve the lye. Use this cup of water to make a very strong tea from fresh mint leaves. Add this back to the dissolved lye mixture, before adding it to the tallow. Check temperature of lye liquid after adding the mint.

Deodorant soap without chemicals: You can use up to two ounces of Vitamin E oil in your soap recipe, adding it to the mixture after stirring in the lye. It has a mild deodorizing quality and is an antioxidant, which will prevent any slightly bacony odor if you have used bacon fat along with your tallow.

Honey complexion soap: Add one ounce of honey and stir it slowly into the soap after adding the lye, and before pouring the mixture into the molds.

Laundry soap: To make laundry soap flakes or powder, let the soap age for three or four days. Grate it on a vegetable grater. Dry the flakes slowly in the oven set at "warm," about 150°, stirring occasionally. It can be pulverized when very dry, or just left in flakes.

Dishwashing jelly soap: Shave one pound of hard soap and boil it up slowly with one gallon of water until it is well dissolved. Put it into covered containers. A handful of this will dissolve fast in hot dishwater. For many soft soap and hard soap recipes and variations, see the interesting Garden Way book, *Making Homemade Soaps & Candles,* by Phyllis Hobson.

MUTTON TALLOW CANDLES

Candles are another good use for the fat that is trimmed off when lamb or mutton is cut and wrapped. While not quite as practical as soap, candles are a fun way to use excess fat and make good gifts, or can be sold.

To prepare candle wicking. Prepared wicking can be purchased (see Sources chapter) but it is simple to make your own from cotton string. One good soaking solution to use is made from eight tablespoons of borax dissolved with four tablespoons of salt in a quart of water. The wicking string is soaked in this for two or three hours, then hung out to dry. Some oldtime candle makers soaked the wicking in apple cider vinegar, or turpentine, and let it dry.

To prepare mutton tallow. Cut up chunks of mutton or lamb fat, put it in a large kettle, and fry it slowly over low heat, as you would for soap. Skim off the bits of fat as they rise to the top. Stir occasionally and do not rush the process and burn the fat. A large batch will take several hours to render out. When it is all pretty well melted down, strain it through a cloth.

Purify it. In a large kettle, dissolve five pounds of alum in ten quarts of water, by simmering. Add the tallow, stir and simmer about an hour, skimming. This not only purifies the tallow, but makes it a little harder texture for use in candles. Cool the tallow until you can touch it comfortably, then strain it through a cloth and set it aside to cool and harden. When it is hard, lift it off the water and scrape off the impure layer on the bottom.

This purified tallow can be stored in a cool place for a week or so until you are ready to make candles, or can be refrigerated or frozen.

Tallow burns with a less pleasant smell than wax or paraffin. It can be perfumed by adding a few drops of pine oil or perfume while the tallow is melted, before dipping or molding the candles.

Candle dipping. Melt the purified tallow and pour it into a wide-mouth jar or container that you can stand in hot water to keep the tallow liquid. Next to this container, have another one filled with very cold water, standing in a pan of crushed ice or ice cubes to keep it cold. Since tallow candles have a tendency to droop in hot weather, don't make your dipped candles too long.

Cut a wick about six inches longer than you want the candle to be, and tie one end of the wick to a small stick. If your containers are large enough, you can tie on several wicks, and dip these all at once.

Dip the wick first into the hot tallow, then withdraw it and let it air-harden for a minute. Then dip it in and out of the ice water, which hardens it. Let it drip thoroughly. Keep repeating this process. To make a tapered candle, do not dip all the way to the top each time you dip it. Since each single dip into the tallow deposits such a thin layer on the candle, it takes a lot of dippings.

Molded candles. It is quicker to mold candles. For candle molds, use plastic or paper cups or cut-down milk cartons. They can be sprayed with a non-stick baking spray (the lecithin based type) to keep the candles from sticking to the mold, or just brushed with cooking oil. Metal molds should be both oiled and chilled before the tallow is poured in. There are silicone type preparations that are also used for "mold release." As with dipped candles, a shorter and wider shape is best when using tallow, which is not as firm as a wax candle.

Since the bottom of the mold will be the top of the candle, ideally the wick should be threaded out through the bottom, and protrude about an inch. This is easily done when using paper or plastic containers for molds. If you can't make a hole in the bottom of your mold, leave a little coil of extra wick in the bottom that you can pull out when the candle is removed. If you have a wick sticking out the bottom of the mold, you can knot it there so you can pull it straight and tight while pouring in the tallow. It could be fastened at the top to a wire or stick that rests on top of the mold. This would keep the wick straight and centered in the candle until it hardens.

Colored candles. Stir in two teaspoons of powdered household dye, like Rit or Diamond Dye, for each pound of tallow, and mix well into the liquid tallow.

SHEEP CARPENTRY

This book has plans for building various pieces of sheep equipment. There is always a need for useful equipment, and you may find a ready sale for duplicates of the pieces you make for your own use.

WASTE WOOL FOR HOME INSULATION

This is an offbeat use for belly wool and skirtings, if you are removing these before selling your fleeces to spinners.

Wash the junk wool well in detergent, rinse it, and squeeze out the rinse water. Treat it with the following solution, suggested by Dr. L.F. Story of the Technical Service of the Wool Organization in New Zealand. It makes the wool both mothproof and fireproof.

1 to 2 pounds sodium fluoride (poison, handle with care)
4 pounds borax
2 pounds boric acid
Mixed into 10 gallons of water

Stir the wool in the solution, squeeze out the water, and dry it without rinsing it. Dispose of the surplus fluid where it will not contaminate pastures or a water supply. Wash your utensils and your hands well.

When the wool is dry, tease it or card it, and spread it evenly between studs, without leaving air gaps. It may be covered with aluminum foil or heavy wrapping paper, if desired.

WOOL SALES TO HANDSPINNERS

The great interest in handspinning has recently created a specialty market for good fleeces. If you keep the fleeces clean, relatively free of grain, hay, burrs and other vegetable matter, have them well sheared (few second-cuts) and handled carefully after shearing, you probably have a product that is valuable for handcraft use.

The fleeces most in demand are the black sheep fleeces, but there is also a good market for clean white fleeces to spin white yarn and to spin yarn for vegetable dyeing. To find the buyers who need these fleeces, contact the nearest place teaching spinning and weaving, and leave your name with them or talk to their students.

Corriedale breeding ewes on the twenty-acre farm of Oldebrooke Spinnery in Lebanon, NJ. Their wool is raised for selling to handspinners.

If you are on a well-traveled road, try putting up a sign saying:

RAW WOOL AVAILABLE FOR HANDSPINNERS

When you get a following of spinning customers, you won't need the sign. If they are happy with your wool, they will be regular customers and tell their friends. If you are in a remote place with little traffic, try advertising in craft magazines such as:

Black Sheep Newsletter
2806 Ham Road
Eugene, OR 97405

Shuttle, Spindle and Dye Pot
Box 7-374
West Hartford, CT 06107

Weavers Newsletter
P. O. Box 259
Homer, NY 13077

Interweave
2938 North County Road 13
Loveland, CO 80537

MORE MONEY IN BLACK FLEECES

Did you know that black lambs have black tongues? The poet Virgil (70–19 B.C.) advised sheep breeders to choose rams lacking pigment in their tongues, to keep fleeces white. In many breeds, the dark or part-dark, vari-colored tongue indicates a white sheep with recessive dark genes. That sheep will be valuable in a black-sheep breeding program, if its fleece is the kind you want.

The "black sheep" of the sheep family is the odd dark lamb that can crop up occasionally in almost any white breed, the result of recessive genes. In large herds a black sheep is undesirable. Its fleece must be handled and sacked separately. Even in the flock, its black fibers may rub off on the burrs in fleeces of white sheep, causing them to be discounted in price because of the special problems caused later in the manufacturing processes.

For handcraft use, the picture is different. The last few years have seen a tremendous interest in handspinning, with many weavers and knitters spinning yarn for their own use, and some spinning for sale. This has created a new market for fleeces, and those most desired are the black sheep ones. In 1974, *Shepherd* magazine wrote, "the unwelcome black sheep has suddenly become respectable, with its wool bringing up to several times the price of white wool."

In areas where there are spinning classes or groups, sheep breeders are discovering that dark fleeces command a high price. The need for dark fleeces exceeds the supply, and although many more people are raising them, the craft classes are turning out spinners faster than the supply of wool is growing.

This interest in dark sheep is appearing in other parts of the world, too. *The Wool Record,* writing about Australia in November of 1975, said, in an article entitled "Production of Black Sheep Being Hurriedly Expanded,"

"The revival of handspinning and weaving has caused a reversal of the century-old effort to eliminate black lambs among Australia's 150 million sheep. The market for normal white wool is depressed but naturally black or coloured wool is undergoing a boom because of the upsurge of interest in handspinning. Craftsmen prefer naturally black and coloured wool to dyed wool for hand-spinning, weaving and knitting. There is a shortage of this wool because of the decades of attempting to eliminate coloured sheep and to produce only white wool."

The specialty breeders of black sheep there have flocks that seldom exceed a few dozen sheep, in a country where flocks of 5,000 are common, but they have most of their fleeces spoken for in advance. In this country, too, most black sheep raisers have small farm flocks, either using their wool or selling it to other handspinners, and selling breeding stock.

In the January 1976 issue of *Shepherd* magazine, the Sheep Buyers' Guide listed more black sheep breeders than those of eighteen other breeds that were listed. The only breeds to outlist them were Corriedales, Dorsets, Finnsheep, Hampshires, and Suffolks.

The fleece of most black sheep will tend to lighten, from year to year. In the beginning it may be a disappointment, but in the long run it will prove an advantage, because it gives a greater variation in color from a relatively small flock. So, in shopping for a black lamb, remember that however black she is at birth, she seldom stays jet-black, but lightens every year. Don't consider the degree of darkness as the main factor in your selec-

Black lambs at the author's home.

tion. Look at her body type and wool grade, which does not change in her lifetime, and probably will be inherited by her offspring.

With one or two black sheep to introduce the black genes, you can in a few years work toward the development of a flock of dark sheep. The easiest way to do this is by the use of a good black ram, of a nice wool breed for spinning. While there is a great variety in the type of wools preferred by various spinners, the majority wants a medium grade wool, with good length for that particular grade. In some parts of the country, preference is influenced by the limitations of a locally made spinning wheel that is popular there. Where there are many spinning classes, the opinions of the teachers are communicated to the students. If there are such classes nearby, ask the teachers what breed of wool they most favor.

Backcrossing for black. Once you get a black ram of a suitable wool type, use him to breed a small flock of white ewes. Geneticists say that the offspring will be "white, but carriers of the black gene." However, in practice we have had people use one of our dark rams on their white ewes and more often than not they got dark lambs.

The first generation of this cross is called the "first filial generation," or F_1. If the F_1 ewes are bred back to the original black ram, their father, this is called a backcross. It produces a generation of F_2 lambs, and in theory there should be as many black lambs as white ones, with all carrying the recessive black gene. You can produce quite a flock in a few years by continuing to breed the original F_1 ewes to the black ram each year. This much inbreeding is not considered to have a great chance of breeding defects, but it is risky to continue it with succeeding generations (like the F_2 offspring). By the time you get a good number of your ram's granddaughters (the black F_2 ewes), you would do well to sell the original ram and get a different one, not related to your sheep.

In raising wool for handspinners, the proof of your success will be repeat customers. This depends not as much on the breed of sheep or the wool type as it does on fleece condition. If the wool is poorly sheared, or full of burrs and seeds and other vegetable matter that they must pick out by hand, or if you have not discarded the heavy dung tags, you may sell someone the first fleece, but you're not likely to sell them a second one. For top prices, you might even consider using sheep coats on your sheep (details on this in the chapter on wool).

The selling of dark fleeces alone is not going to support a flock of sheep in the way they would like to become accustomed. But, you can add to this the selling of black sheep breeding stock, and the sale of locker lambs, and have one in your own freezer.

To get an even better price per pound, consider the actual handspinning and selling of yarn.

HANDSPUN YARN

Spinning your own wool into yarn is one way to compound its value per ounce. You can spin it for your own use in knitting, or practice on it for home use and go on to sell it to other knitters when you get good enough.

In the Appendix there is a list of spinning literature and sources for spinning wheel plans to build your own wheel. See my *Spinning and Weaving with Wool* (Appendix).

INCENTIVE PAYMENTS FOR WOOL SOLD

The county Agricultural Extension agent can tell you the current status of incentive payments based on your raw wool sales, as this changes from year to year.

Incentive payments are made in accordance with the National Wool Act, as amended by the Agricultural Act of 1970 and the Agriculture and Consumer Protection Act of 1973, and are price support payments. The "shorn wool payments" are based on a percentage of each producer's returns from the wool he has sold. The percentage will be that required to raise the national average price received by all producers of shorn wool, up to the announced incentive price. The 1976 incentive was 72¢ a pound, meaning that if the national average price received for wool was less than 72¢ a pound in 1975, an incentive payment would be made to all the people who had sold wool and filed the proper forms. Since the 1975 average price did prove to be much less than 72¢ per pound, the incentive payment was computed on the average and announced to be 61.1 percent of the net proceeds, for everyone who had filed.*

This is not funded directly from tax revenues. A tariff charged on all lamb and mutton imported into this country is used to fund the wool incentive payment program.

The Agricultural Extension Service usually has the forms that must be filled out, although the actual department in charge of the payment is the Agricultural Stabilization and Conservation Service. You must file prior to January 31 for payment on wool sold the previous year, and after that the percentage is computed, and the payments are made in about April, for the previous year's sales. If you use your own wool rather than sell it, you are ineligible.

As an example, the 61.1 percent of the net proceeds means that if you had 100 pounds of wool and sold it for 50¢ a pound, you would be paid 61.1 percent of the $50 that you received for your wool, $30.55 in incentive payment. But, on the other hand, if you sold your 100 pounds of wool for $1 a pound by keeping the fleece clean and careful shearing and then locating a handspinner who appreciated the care you had taken, you would have received $100 for the wool you sold, making your incentive payment $61.10.

This creates a real "incentive" for you to sell your wool for the best price you can, per pound.

---

*Percentage fluctuates from year to year. In 1974 it was only 21.8 percent. No checks are made out and sent for an amount under $4 unless the producer requests it.

CHAPTER TWENTY

SHEEP CALENDAR

The calendar starts with the purchasing of ewes and/or ewes with lambs in May, as this is the month of most purchases of beginning flocks. (Replacement ewes are more often bought in August-September.) Many people move to the country in late summer, don't get fenced and organized for sheep till spring, and want to buy them when they see all that new pasture coming up.

MAY

IF YOU JUST PURCHASED SHEEP:

Check your fences before bringing home your sheep. It is easier to keep sheep in than to try to get them back in. Fences are also to keep dogs out, so have them as dog-proof as possible.

Build a corral that makes it quick and easy for you to round up sheep whenever necessary.

Check your pasture for presence of toxic plants. If you need help in identifying them, contact your county agricultural extension agent or a helpful neighbor or sheepraiser.

Don't turn the new sheep onto more lush pasture than they had before, without giving them hay first. If pasture is extremely lush, turn them out on it for only part of the day, for the first two days.

Check for ticks before you turn your sheep loose. Treat them if you see even *one*.

Check with neighbors who own dogs that normally frequent the area. Let them know you now have sheep.

IF YOU ALREADY HAVE SHEEP:

Slosh or spray for ticks (Chapter 14) if you had heavy infestation when you treated them after shearing, or if there is any sign of ticks since then.

This is a good month to take a little vacation. Arrange with a neighbor to keep an eye on things and watch for dogs, and replenish feed and water in the creep, and water for the mature sheep.

Sunday	Monday	Tuesday	Wednesday	Thursday	Friday	Saturday
				1	2	3
4	5	6	7	8	9	10
11	12	13	14	15	16	17
18	19	20	21	22	23	24
25	26	27	28	29	30	31

JUNE

Sunday	Monday	Tuesday	Wednesday	Thursday	Friday	Saturday
1	2	3	4	5	6	7
8	9	10	11	12	13	14
15	16	17	18	19	20	21
22	23	24	25	26	27	28
29	30					

Your pasture is best grazed at 4-6 inches high. Rotate pastures to control grass length and let it regrow, also to help control parasites.

Locate roofed salt boxes on pasture near clean water. See Chapter 13.

Watch for fly strikes and maggots. Fly strike on sheep is just a cluster of tiny eggs at first, and you have to look closely to detect it. When maggots first hatch, they are very small but still easy to see. Investigate any damp spots on fleece, or excessive rubbing on fences, for signs of maggots (see Chapter 14). Cut away the wool from any infested spot and treat it with screwworm spray (Sources chapter).

Messy rear ends invite flies, so keep them trimmed if possible. Any wound or dog bite is also attractive to flies. During hot weather you will have to be alert to the maggot danger.

JULY

Sunday	Monday	Tuesday	Wednesday	Thursday	Friday	Saturday
		1	2	3	4	5
6	7	8	9	10	11	12
13	14	15	16	17	18	19
20	21	22	23	24	25	26
27	28	29	30	31		

Keep flock cool if weather is extra hot. Make some shade arrangement if you have no trees. Keep ram near shade or in barn if over 90°. Heat adversely effects virility. Sheared scrotum will help keep him cool.

Provide cool fresh water for all the sheep. Grown sheep need one to two gallons per day.

Put your ram in shaded pasture next to the ewes now, if you want to start breeding them in August. His presence nearby helps bring them into heat a little sooner. Start flushing ewes (see Chapter 6) seventeen days before you plan to turn the ram in for breeding.

Watch for bloat if they are on lush pasture. Do not pasture ewes on clover now, before breeding time. It has a slight birth-control effect on ewes.

Watch for limpers, check feet, trim them and foot-bathe if necessary.

An orchard is ideal shady place for sheep in summer. But don't let them get too much fruit, or too suddenly. Overeating fruit can cause bloat and death. See Chapter 3 for protection of tree trunks from sheep damage.

Worm the sheep. See Chapter 13 for worming directions.

Mark down worming date, check against that date if selling locker lambs.

If you send any lambs to slaughter, ask for the return of the pelt for tanning. It is a good source of extra income.

Sunday	Monday	Tuesday	Wednesday	Thursday	Friday	Saturday
						2
3	4	5	6	7	8	9
10	11	12	13	14	15	16
17	18	19	20	21	22	23
24 31	25	26	27	28	29	30

Provide access to shade. Sheep can get sunstroke in excessive heat, and rams become infertile if temperature is over 90° for long. Keep ram cool by reshearing scrotum, keep in shady place in daytime, turn in with ewes in evenings and at night, breed in August for early lambs. Ram can service more ewes, if used only in cool evenings and night.

If you are going to cull out any ewes, do it before breeding. Cull gummers, broken mouths, unproductive ewes. Keep good milkers, early lambers, twinners. Consider putting a cull in your own freezer. Mutton is very digestible, and very good burger meat (see Chapter 18 for muttonburger recipes).

Do mid-season worming, if you didn't do it in July.

Keep ewes off clover pasture prior to breeding, since it can suppress mating hormones. Rams need small amount of grain daily; ¼ to ½ pound will keep them in shape.

If you share a ram with a neighbor, be sure to keep him with ewes six weeks at least, so there is a minimum of two chances for mating. Mark exact time on calendar when you put ram with ewes, so you know what date is earliest day you can expect lambing to start.

Keep record of dates of marked ewes, if you use a marking harness (Chapter 5). That way you can separate early bred ewes for pre-lambing ration, and know when to watch for lambs.

Pick up grain for winter if price is right. Store in rodent-proof drums, where sheep cannot break into it. Overeating grain can be fatal. If your pasture includes wooded area, count sheep daily, check for down ewes or ones caught in a fence.

SEPTEMBER

Sunday	Monday	Tuesday	Wednesday	Thursday	Friday	Saturday
	1 Locate your winter supply of hay.	2 On ewes you think are not yet bred, trim wool at rear end if it has accumulation of manure tags, to make it easier to breed.	3	4	5	6
7 Keep water and salt always available.	8	9 Make list of needed repairs and start before winter.	10	11	12	13
14 Worm any late-born lambs still to be fattened to market size, if it is thirty days or more before slaughter.	15	16 Get sizeable lambs to slaughter, sheared first. If too short to shear, ask for pelt back, for tanning.	17 If you are breeding for late lambs, start flushing ewes for seventeen days, then turn in the ram.	18	19	20
21 Record dates of any witness of breedings; write it on calendar so you know when to watch for their lambing.	22 Reduce grain ration gradually after flushing. Then discontinue grain for overfat ewes to thin them down so you can put them on rising plane of nutrition in last four–five weeks of pregnancy.	23	24	25	26	27
28 Clean out barn before you get your winter hay, spread manure on parts of vegetable garden already harvested.	29 If you have orchard, feed some (not too many at a time) windfalls to the sheep.	30 Set some windfalls aside to feed later. Sheep can use those vitamins in winter.				

217

OCTOBER

Sunday	Monday	Tuesday	Wednesday	Thursday	Friday	Saturday
			1	2	3	4
5	6	7	8	9	10	11
12	13	14	15	16	17	18
19	20	21	22	23	24	25
26	27	28	29	30	31	

Clean out sheep sheds and barn, if not done already. Spread on garden.

Get in your winter hay, and get barn ready for winter.

Clean out and check waterers, winter-proof your faucets.

If there has been no visible sign of breeding yet, get a loaner ram, quick. You may not notice a lot, but you should have seen him really cozy with a ewe, for a day or two, following her everywhere.

Check over lambing supplies, if you order by mail. See Sources chapter for mail order supply houses.

Get lamb pens made, if you need them. (See Chapter 9.)

218

NOVEMBER

Sunday	Monday	Tuesday	Wednesday	Thursday	Friday	Saturday
						1
2	3	4	5	6	7	8
9	10	11	12	13	14	15
16	17	18	19	20	21	22
23/30	24	25	26	27	28	29

Rid your feed room of rodents and use rodent-proof containers. Grain is too expensive to feed the rats.

Check lambing supply inventory; order if not already done. Order ear tags now, if you plan to use them to identify lambs. See Sources chapter.

Order antibiotics and store in refrigerator for emergencies.

Separate out any limping ewes, trim feet and medicate.

Remove ram if you have not already, to protect pregnant ewes from injury.

If ram is run down, feed him well to restore condition.

Add stock molasses to drinking water for pregnant ewes. They need the sugar. This is less messy than adding to feed.

Start preparing lambing pens if you don't have folding ones that can be set up as needed. See equipment plans for folding lamb pens.

Sunday	Monday	Tuesday	Wednesday	Thursday	Friday	Saturday
	1	2	DECEMBER	4	5	6
Crotch ewes to prepare for lambing, handling them carefully. See Chapter 7 for crotching. Remove dirty tags from udders and legs.						
7	8	9	10	11	12	13
Separate out obviously advanced pregnant ewes. Feel their udders; if not soft, they may have no milk for lamb. Be prepared to feed it. Put molasses in drinking water for ewes. It is good for them and also helps keep water from freezing.						
14	15	16	17	18	19	20
Feed both grain and hay to advanced pregnant ewes (within four weeks of lambing): ¼ to ½ pound of grain is not too much. It is better to feed more grain and less hay at this point. They don't have room inside for lots of bulky feed, and too much hay can help bring on prolapse. If available, add up to 40 percent wheat (for its selenium content) to grain mix for pregnant ewes.						
21	22	23	24	25	26	27
Watch for pregnancy toxemia in ewes near lambing, if listless acting, and treat with propylene glycol. See Chapter 12.						
28	29	30	31			
Deficiency of calcium can predispose ewe to hypocalcemia or cause deformed teeth in her lambs as they grow and poor fleeces, so give supplementary source of calcium. See Chapter 13.						

JANUARY
(or if late lambing, do this the month of lambing)

Sunday	Monday	Tuesday	Wednesday	Thursday	Friday	Saturday

Be sure pregnant ewes are getting exercise.

Maintain regularity of feeding schedule and amount, and notice actions of ewes at feeding time. Refusal to eat is one sign of pregnancy toxemia (Chapter 12), or being just about to lamb. If they are listless or staggering, it is likely pregnancy toxemia.

Add molasses to ewes' drinking water, as this helps in preventing toxemia.

Crotch ewes, if you don't shear them before lambing, and did not crotch last month. Be vigilant for ewes needing help while lambing. See Chapter 7 for normal procedure; see Chapter 8 for difficult lambings.

Be prepared to lamb if power fails in winter storms. Have lantern ready, and hot water bottle or plastic jug of water, instead of heat lamp.

Check barn electrical connections for heat lamp use. Sockets for bulb must be ceramic, to be safe.

Make rounds regularly at lambing time, to spot ewes going into labor, until all have lambed.

Immediately after ewe lambs, squeeze her teats to get them unplugged, put iodine on lamb's navel. Watch to see that it nurses; help if necessary. Give ewe warm molasses water to drink. Pen them together for the first few days. (See Chapter 7.)

Dock and castrate lambs at three to ten days old. See Chapter 9.

Check lambing area for booby traps: holes for lambs to climb in among bales of hay, hurdles that could fall on them, heat lamps that could fall and cause fire.

FEBRUARY

Sunday	Monday	Tuesday	Wednesday	Thursday	Friday	Saturday
						1
2	3	4	5	6	7	

Pen newborn lambs with their mothers for two to three days, with water and hay for ewe. Lamb pens and hay racks are shown in Chapter 4.

Watch newborn lambs, especially first three days, to be sure their rear end is not plugged up, or runny. See Lamb Problems, Chapter 11.

| 8 | 9 | 10 | 11 | 12 | 13 | 14 |

Watch twin lambs. The slower growing one may benefit from a supplementary feeding of lamb milk replacer, in bottle.

Milk out any ewe who has lost her lamb, or get her to take an orphan (Chapter 10) to prevent mastitis. Freeze colostrum in small plastic bags or ice cube trays, for easy defrosting if needed. Milk her on third day, sixth day, and once more if needed.

| 15 | 16 | 17 | 18 | 19 | 20 | 21 |

Dock lambs tails, and ear tag them if you use tags, at three days old, before turning out ewe and lambs. You won't believe how hard it is to catch a little lamb, once it is turned out of lambing pen.

Check ewe's feet, trim and treat if necessary, before turning out of pen. If suspicious of hoof disease, treat hoof after trimming, see Chapter 15.

| 22 | 23 | 24 | 25 | 26 | 27 | 28 |

Give plenty of fresh clean water. Ewes need it to make milk. Ewe with lamb needs 1½ gallons per day, dry ewe 1 gallon, lambs ½ gallon. Plentiful water encourages lambs to eat better, and helps prevent urinary calculi in ram lambs.

Give trace-mineral salt, free choice. If you have hard water, this is more important. Salt makes them thirsty, so they drink more water, for good health.

| 29 | | | | | | |

Continue to grain lactating ewes to keep up milk supply.

Prepare lamb creep in a good location: dry, away from drafts, well lighted because lambs will eat better, and near where the mothers congregate.

MARCH

Sunday	Monday	Tuesday	Wednesday	Thursday	Friday	Saturday
1	2	3	4 (Creep plans, Chapter 11.)		5	6
7	8	9	10	11	12	13
14	15	16	17	18	19	20
21	22	23	24	25	26	27
28	29	30	31			

Get lambs on creep feed as soon as possible. (Creep plans, Chapter 11.)

Check docking and castration bands for any sign of infection. If any, clean out well, douse with strong iodine. If you keep horses, or are in tetanus area, you should have vaccinated the lambs.

Watch for lambs not growing well, and give special attention as needed. Look for pale inner eyelids or other sign of worms.

Worm all sheep with broad-spectrum wormer such as *Tramisol*, *Loxon*, or *Thibenzole* (not with copper sulfate or pheno-thiazine right now before spring. Grass has so much copper now, and excess copper is toxic).

Avoid giving trace mineral salt now in the spring, for the above reason.

Restock worming supplies, if you are low on them.

Clean out barn, spread winter's accumulation of manure and bedding on your vegetable garden plot (really makes it grow) or on area of pasture that needs it.

Start shearing if weather permits, and you have not done it yet. See Shearing Chapter 17.

When shearing, try to keep ewes and their lambs together as much as possible, to avoid confusion and disowning.

Examine udders when shearing, mark ewes for culling if they have damaged teats or hard udders. Consider culling any who did not breed.

Worm sheep and trim hooves at shearing time, if not already wormed. Do not shear sheep if they are wet.

Keep fleeces clean by sweeping floor or shaking out tarp after each sheep.

At shearing, or ten days later, treat for ticks (see Chapter 14). Slosh, spray, powder, or sprinkle.

Try to get the best price for your wool. If it is nice and clean and well sheared, handspinners may be your best customers. Keep receipts for wool sold, to apply for incentive payment (see Chapter 19).

APRIL

Sunday	Monday	Tuesday	Wednesday	Thursday	Friday	Saturday
		1	2	3	4	5
6	7	8	9	10	11	12
13	14	15	16	17	18	19
20	21	22	23	24	25	26
27	28	29	30			

1. Feed on well drained ground to avoid hoof troubles. 2. A portable feeding arrangement (feeders, Chapter 4) keeps them from always trampling same area.

7. Keep fresh water, salt, and feed in your lamb creep.

8. Before turning ewes into any new pasture, let lambs in first. This is "advanced creep feeding," and gives them the best of the grass.

13. Worm lambs at about forty pounds, if not done. They can be weaned then if you have good separate pasture for them. Put feed directly into lambs as ewe milk production declines. 14. If you wean lambs, dry off ewes on hay ration alone to make sure milk dries up, to avoid mastitis.

20. Avoid trace mineral salt when you have first spring grass, for grass is already unusually rich in copper. 21. Make covered pheno-salt box for pasture (plans, Chapter 13). Add calcium for the lambs.

27. Did you clean out the barn or shed accumulation and apply it to your garden plot? If you don't need it, try selling to a gardener, or trade for some of their vegetables.

28. Consider for culling: Ewes with bad udders, teeth, hooves, or ones who did not do a good job of raising lambs, or ones who did not lamb (through no fault of yours) and ones with poor fleece.

APPENDIX A

VETERINARY & SHEEP SUPPLIES

Source	*Products*	
Sheepman Supply Co. P. O. Box 100 Barboursville, VA 22923	Worming boluses Injectable Tramisol Tetanus antitoxin Vaccines Sheep dip for ticks Foot rot spray Formaldehyde for hooves Caldex MP for toxemia Caldex MP for milk fever 7 percent iodine Disposable syringes	Elastrators and bands Prolapse retainers Sheep shears Shepherd's crooks Lamb milk replacers Lamb nipples Lam-Bar & nipples Ear tags Disposable lamb coats Lamb-reviver stomach tube BO-SE for white muscle disease of lambs
Nasco Farm and Ranch 901 Janesville Ave. Fort Atkinson, WI 53538 or Nasco West 1524 Princeton Ave. Modesto, CA 95352	Ear tags Sheep shears Shepherd's crooks Lambing snares Burdizzo hoof shear Dose syringes Medications Balling guns Sheep coats Fence supplies Fence stretchers	Barbed wire dollies Septi-lube, antiseptic lubricant Capsule forceps for boluses Foot rot spray, for all hoof problems
Kansas City Vaccine Co. Stock Yards Kansas City, MO 64102	Bacterial scour treatment (for lambs) Bloat medications, several kinds Sulfon-o-mix for pneu- monia & lamb scours Sheep worming tablets Strong iodine Emasculatomes	Pheno-lead arsenate boluses Keto-stix (ketosis test strips) Hypo-Nitrit solution (for treatment of arsenic or lead or prussic acid poisoning)

Source	*Products*
V-E-T-S Company 203 West Avenue U. Drawer C Temple, TX 76501	Tetanus antitoxin 50 percent dextrose solution Sterile disposable syringes & needles *Merck Veterinary Manual* (very technical) Blood stopper with insect repellent Screwworm and ear tick spray Lamb nipples
C. H. Dana Co., Inc. Ranch Supplies Hyde Park, VT 05655	Sheep bells Sheep coats and blankets Capsule forceps Hay hook Barbed wire stretcher Ear tags
Dakota Stockman's Supply Box 12 Brookings, SD 57006	Veterinary thermometer KRS maggot-killer bomb Capsule forceps Wound-kote bomb Kopertox
C. J. Martin Co. P. O. Box 1089 Nacogdoches, TX 75961	Foot rot liquid Amine-iodide
P. A. Gardner Ltd. P. O. Box 4237 Auckland, 1 New Zealand	Soft veterinary fingertips for lambing-loop $1.65 including postage. I give price on this because of high cost of postage for letters going back and forth to inquire about price, and another letter necessary to order it and send payment.
Clyde Thate Route 1, Box 29 Burkett, TX 76828	Send stamped envelope for information and price of his coyote snare
Mid-States Wool Growers Coop 3900 Groves Road Columbus, OH 43227	#2 Bronze sheep bells, for dog control Lead-arsenate worm boluses Vaccines
Creutzburg, Inc. Box 7 Lincoln Hwy. East Paradise, PA 17562	New fence-crimping tool called "Fence Tight" (great!)

FOR SPECIFIC SUPPLIES

Tanning Supplies:

25% comm'l glutaraldehyde solution
Eugene Brown
128 Green Ave.
Hemstead, NY 11550

Glutaraldehyde:

H and H Enterprises
501 20th Avenue
Anchorage, AK 99503

Pelt Tanning Kits:

U-Tan Co.
P.O. Box 4014
Warren, NJ 07060

Tanning done to order:

Bucks County Fur Products
P. O. Box 204
Quakertown, PA 18951
Send stamped envelope when inquiring
    for price.

Meat curing kits, sausage casings, seasonings, grinder, stuffer, etc.:

RAK, National Home Products
P. O. Box 4397
Las Vegas, NV 89106

Sausage seasonings, casings, etc.:

Sedro Industries
P. O. Box 8009
Rochester, NY 14606

Softal tanning oil:

J. W. Elwood Supply Co.
Omaha, NE 68102

Candle supplies, molds, fragrances, wicking, etc.:

Vermont Village Shops
Bennington, VT 05201

Spinning wheel plans, 25¢ catalog of Craftplans

Craftplans
Industrial Boulevard
Rogers, MN 55374
Has 3 styles of wheel plans.

APPENDIX B

SHEEP PERIODICALS, BOOKLETS, AND BOOKS

PERIODICALS

Shepherd magazine (the best)
Sheffield, MA 01257
(Currently $4.95 a year.)

The Wool Sack (small newspaper)
Drawer 238
Brookings, SD 57006
($1.00 a year.)

National Wool Grower
   (useful, but for large sheep raisers)
600 Crandall Building
10 West Stars Avenue
Salt Lake City, UT 84101
(Currently $5.00 a year.)

Sheep Breeder and Sheepman
P. O. Box 796
Columbia, MO 65201
(Currently $5.00 a year.)

Wyoming Wool Grower
Wyoming Wool Growers Assoc.
P. O. Box 115
Casper, WY 82601
(Included in membership to
Wyoming Wool Growers.)

Montana Woolgrower
P. O. Box 1693
Helena, MT 59601
(Subscription included in
Montana Woolgrowers Assoc.
membership.)

Rural Research (quarterly)
CSIRO Publications Service
P. O. Box 89
East Melbourne, Victoria
Australia
(Semi-technical agricultural research
magazine, currently $4.00 a year.)

Black Sheep Newsletter
28068 Ham Rd.
Eugene, OR 97405
($4.00 a year; 80¢ each for back copies.)

The Ranch magazine
   (sheep, goats, cattle)
P.O. Box 1840
San Angelo, TX 76901
(Currently $5.00 a year.)

BOOKLETS & BULLETINS

Breeds of Sheep
Jones Sheep Farm
Peabody, KS 66866 $3.50
—booklet with pictures and information on 25 breeds of sheep.

Handspinner's Guide to Selling
by Paula Simmons
Pacific Search Press
222 Dexter Avenue N.
Seattle, WA 98109
—experience-filled book for solving
and avoiding problems in selling.
$10.95 postpaid

Patterns for Handspun
Paula Simmons
Box 12
Suquamish, WA 98392
—over 30 knit and crochet patterns
for use with your handspun wool yarn.
$1.75 postpaid

*Plans for Farm Buildings and
    Equipment*
Extension Service
Utah State University
Logan, UT 84321
—a catalog showing many plans for
which they can supply drawings.
$1.25 postpaid

*Sheep Handbook of Housing and
    Equipment*
Midwest Plan Service
207 Agr. Engr. Bldg.
Iowa State University
Ames, IA 50010
—detailed working drawings for sheep
equipment.
$5.00 postpaid

Sheepman's Production Handbook
American Sheep Council
200 Clayton Street
Denver, CO 80206 $6.95
—for medium and large herds.

Spinning & weaving articles, reprinted
from *Handweaver* magazine. Send
$2.50 to:
Paula Simmons
Box 12
Suquamish, WA 98392

Spinning for Softness and Speed
by Paula Simmons
Pacific Search Press
222 Dexter Avenue N.
Seattle, WA 98109

Spinning and Weaving with Wool
by Paula Simmons
Pacific Search Press
222 Dexter Avenue N.
Seattle, WA 98109
$13.95 postpaid

Tanning Instructions for Sheepskins
Shepherd magazine
Sheffield, MA 01257
—a tanning reprint package which
shows several methods, including
USDA's glutaraldehyde method.
$1.95 postpaid

Warp and Weft reprint articles
Paula Simmons
Box 12
Suquamish, WA 98392
—useful spinning, weaving and sheep
articles.
$1.95 postpaid

INDEX